CONQUERING CARBON

CONQUERING CARBON

CARBON EMISSIONS, CARBON MARKETS AND THE CONSUMER

FELICIA JACKSON

Contents

Acknowledgements — 07

Foreword by Lucy P. Marcus — 08

Introduction — 11

1 **Climate Change and Greenhouse Gases** — 15

Why conquering carbon is critical
Why do we need a carbon price?
Driving behavioural change
Transition to a low carbon economy

2 **The Resource Wars** — 33

Living beyond our means
Extreme weather events and social disruption
Conflict and climate change
Oil and the economy
Food security
What about water?
The real cost of carbon

3 **Driving Change** — 55

Risk versus opportunity
A green 'New Deal'
Changing behaviour
Setting a price on carbon
Measure to manage

4 **Decarbonizing our world** 79

Decarbonizing power: stabilizing emissions;
decarbonizing power sources; transforming
infrastructure
Beyond power generation: efficiency; IT; buildings;
transportation; forestry

5 **The International Policy Framework** 151

The Kyoto Protocol
EU Emissions Trading Scheme
Voluntary versus compliance markets
Reaching an equitable agreement
Alternatives approaches
Achieving a post-Kyoto agreement

6 **What We Can Do** 175

Countries
• Selected country programmes: United Kingdom,
United States, China
Cities
• Selected city programmes: São Paolo; San Francisco;
Seattle; Rizhao
Corporations
• Selected coroporate activity: Wal-Mart; Tetra-Pak;
Google; The Savoy Hotel
Consumers

Conclusion 233

Further Reading 240
Acronyms and abbreviations 247
Glossary 248
Index 254

First published in 2009 by New Holland Publishers (UK) Ltd
London • Cape Town • Sydney • Auckland
www.newhollandpublishers.com

10 9 8 7 6 5 4 3 2 1

Garfield House, 86–88 Edgware Road, London W2 2EA, UK
80 McKenzie Street, Cape Town 8001, South Africa
Unit 1, 66 Gibbes Street, Chatswood, NSW 2067, Australia
218 Lake Road, Northcote, Auckland, New Zealand

A catalogue record for this book is available from the British Library

ISBN: 978 1 84773 425 9

Publishing Director: Rosemary Wilkinson
Publisher: Aruna Vasudevan
Project Editor: Julia Shone
Editor: Luisa Moncada
Cover design: Hello Paul Limited
Book design: Rebecca Longworth
Illustrations: Alan Marshall
Production: Melanie Dowland

Reproduction by Pica Digital Pte. Ltd., Singapore
Printed and bound in India by Replika Press

The paper used to produce this book is sourced from sustainable forests.

Acknowledgements

For my parents, Tom and Paula.

This book could never have been written without the support of family and friends, as well as the time and expertise freely given by many in the areas of climate change and carbon, either in interviews, in conversation or in lectures. There were many people who have been of great help but in particular I'd like to thank:

Michael Brown, Tom Burke, Jenny Chase, James Cameron, Celestine Cheong, Henry John Drexhage, Dagmar Droogsma, Samuel Fankhauser, Vanessa Frances, Lionel Fretz, Barry Gardiner, Ed Gillespie, Clare Goodess, James Graham, Chris Greenwood, Mike Grenville, Michael Grubb, Jenny Harrison, Polly Higgins, David Hopkins, Rob Hopkins, Steve Howard, Belinda Howell, Philippa Jackson, Joanna Lee, Richard Lorch, Ewan Murray, Peter Myers, Simon Reddy, Mike Scott, Lisa Silverman, Charlotte Streck, Oliver Tickell, Bruce Usher, Allesandro Vitelli, Tom Whitehouse, Mike Wilson and William Young.

Author's Notes

Currencies have been given in the denomination and amount at the time of announcement. The science, political environment and the financial analysis are all under constant review, so it should be noted that this book is not intended to be an in-depth academic analysis, but rather an overview of the issues at hand: why we need to understand the impact of carbon emissions, how they could affect our world, and what is being and can be done about them. It's a snapshot of our understanding in 2009 and hopefully a tool to help us all become part of the solution moving forward.

While I've included some further reading for more detail, if you want detailed references, further information or just to get in touch, you can reach me at: http://www.conqueringcarbon.com.

Foreword

'Climate change has the potential to change all of our lives in negative ways. If we want to balance the odds in our favour and minimize the chance of rising sea levels, increasing droughts, and difficulties in accessing food, water and health care, we need to act and we need to act now.'

This is the fundamental conclusion that Felicia Jackson reaches in this carefully researched and convincingly argued book. The environmental challenges we already face are enormous, and climate change can only exacerbate them. This is a view that I personally share – climate change, as it exists now, is real, serious, fast-paced and has immediate and long-term consequences. The multiple causes and consequences of climate change, the myriad ways of responding to it, the complexity, yet necessity, of comprehensive public education about it, and its truly global nature make it one of the most profound challenges that mankind has ever faced.

Climate change crosses all borders, requiring concerted action on many fronts: from the political to the scientific, from the economic to the social, from the cultural to the educational. With each perspective comes a distinct agenda regarding the most effective, most appropriate response. In this jungle of opinion, it seems hard, if not impossible, to get a firm grasp on the issues involved. How do we navigate our way through the issues to make up our own minds, about how to contribute as individuals, as well as part of our larger communities, organisations and societies?

To my mind, the most important thing that must be done is to ensure that everyone involved knows what they are talking about. Felicia Jackson makes a difficult task seem easy: in a short space she conveys the main reasons why it is important to control carbon emissions, lays out the various approaches that have been put forward, and examines the problems that beset any attempt to set policy or create international agreement in this area. Though no book can hope to cover all the questions that arise within a subject as complex as climate change, Conquering Carbon takes a giant step towards addressing that knowledge gap and thus allows us to begin to understand the causes and consequences of our own actions past and present.

The book's contribution is particularly welcome because it goes beyond simply raising awareness of the climate change problem and shows how, if we can explain the cost of inaction clearly, we can force economic change. It illustrates concrete ways of what can, and should, be done by all of us individually and collectively. One of the most important benefits of Felicia Jackson's work is that it helps us to clearly understand how the carbon limitation issue is not a distant problem that has been displaced by the current economic crisis. The importance of understanding this problem, whatever the shorter-term economic background, stands out.

The changes required to achieve a low-carbon economy will not come automatically or easily. Governments need to focus less on endless short-term fixes and more on solutions that are sustainable in the long-term. They have a particular responsibility when it comes to choices about energy production and consumption and about a regulatory framework that incentivizes the green economy. But tackling carbon cannot be left to governments alone. Companies, too, are taking the debate on board. Driven by legislation, consumer preference, and their own financial bottom line, they are making better choices.

A regulated carbon market can provide a bridge between the public and the private sectors. For such a market to exist it must be mandated through legislation, and have strong governance and controls. If managed effectively, such an approach can provide a mechanism through which carbon can be cut without decimating existing economic models.

As individuals, we too, have choices to make – choices that empower us, but that also place a responsibility on us – to make it clear how strongly we feel about the clear and present danger of climate change and the need for decisive, comprehensive action now. To borrow from Barack Obama, if we do not have concerted action, we will have collective failure. Above all, we must accept that knowing the facts, educating ourselves where necessary so that we can make informed decisions as consumers, entrepreneurs, investors, policy makers, and parents is not optional. It is a fundamental requirement if we are to succeed in coming to terms with climate change, preventing further damage to our environment, and undoing as much of what has been done as possible.

There is no time to waste. The situation is urgent and we must act now and act decisively. No matter how unsettling the debate, no matter how much fear it may strike in the hearts of even the bravest souls, we have no choice but to address climate change head on, with knowledge and courage and the hope that working together we can achieve our goals. The fact that climate change is a predominantly human-made rather than a natural disaster should underline both our responsibility and the future opportunity. After all it's how we respond, as individuals and as a society, that will define our future.

– Lucy P. Marcus, April 2009

Lucy P. Marcus is the Founder and CEO of Marcus Venture Consulting. Selected as one of the World Economic Forum's Global Leaders for Tomorrow, she is cofounder, judge and Chair of the board of the Aspen Institute Energy and Environment Awards, a member of the international advisory board of the Instituto de Empresa Business School in Madrid, and on the boards of BioCity Nottingham and the International Step-by-Step Association (ISSA).

Introduction

Conquering carbon is the greatest challenge we face in the early 21st century. We need to tackle climate change, to cut the emission of greenhouse gases into our atmosphere, if we are to prevent, or even slow, the advent of global warming.

It's a controversial subject. Not everyone agrees that global warming is a problem and, while awareness of the issues has become more mainstream in the last few years, a great deal of confusion, argument and misunderstanding remain. The underlying science is complex and the analysis ever changing. At the same time, many of those involved in the debates have very specific agendas, from the industrial to the overtly political. This book is an attempt to explain what is actually going on, how we might be affected and how the carbon markets might help us to solve the problem.

Many people have an 'out of sight, out of mind' attitude towards climate change, believing that it's a problem that only affects others. It's certainly true that countries with the least responsibility for emission of greenhouse gases over the last 200 years – the developing economies – are likely to feel its effects most strongly. However, if climate change theories prove accurate, we are facing fundamental changes in supplies of food, fuel and clean water on a global scale. Anyone who thinks of this as someone else's problem is deluding themselves.

Concern about climate change has focused our attention on the need to move away from a dependence on carbon-emitting fossil fuels. This makes finding a way to control our carbon emissions significant to the development of a modern society on two levels: as a step towards removing carbon from our supply chains, as well as a model of how to effectively control and conserve global resources.

Conquering carbon provides us with an opportunity to use climate change policies to address wider global environmental and equitable challenges. How we manage man-made emissions is a test for how we can begin to address issues of access to, and global allocation of, resources in the coming century. For years environmentalists have been warning that we're running out of oil, water, food and even land. With a global population expected to hit 9 billion by 2050, there is no question that we have finite resources that may not be able to serve the expanding global population.

We need to balance supply and demand and sustainability is high on both the political and corporate agenda. The only realistic option that we have is to adapt our models of production and consumption sufficiently to change the way we use our resources. This places us on the cusp of a fundamental transformation of how our economy functions. Successfully addressing the issues of climate change and environmental decline will demand a cultural shift in our expectations of consumption, as well as our expectations of comfort.

This opportunity could potentially transform our economic framework, enshrining the concept of equity and responsibility at the heart of our economy. Whatever route we take, our next steps are going to have an impact on every aspect of modern consumer society, in ways that are likely to accelerate as the consquences of delayed action become ever more clear. We live in a capitalist society and, while the financial disaster of the credit crunch and failing banks may have shown us the flaws of our current economic model, capitalism remains the philosophy that underlies most of the world's economies. That means that we need to take a closer look at the the explicit costs that we're facing – not just the climate changes potentially on the horizon, but the potential cost to all of us if we fail to avert it. Once we accept those costs are real, we can look at methods of using price to change behaviour. By putting a price on carbon we hope to discourage and curb its emission.

The introduction of an international carbon market is one way of doing this. The carbon market has a critical role to play in alleviating the worst effects of climate change; carbon offsets are far more than an opportunity for the developed world to offload their responsibility to cut pollution onto the developing world. For many, these last two statements are even more controversial than any about climate change. The carbon markets were set up as a global public good, part of the first attempt to put a price on the consequences of our actions in the hopes of changing behaviour. Many believe, however, that it has failed to achieve its goals, arguing that alternative approaches have a greater chance of successfully cutting emissions.

The creation of the carbon offset, the reduction or avoidance of greenhouse gases (GHGs) in one place to 'offset' emissions occuring elsewhere, has been accused of providing an opportunity for companies and individuals to offload their responsibility for cutting carbon emissions to the third world. The reality is that the carbon markets are the best hope we have to change the framework of our economy to a low carbon basis and in this book we're going to explore why and how.

The trading of carbon credits is a simple concept in theory. Every emitter of carbon dioxide is given a number of credits equal to an agreed level of emissions. That number of credits, or certificates giving the right to emit one tonne of carbon, must be surrendered to authorities at the end of the year. If they don't have enough, they have to buy more from others who have managed to cut their emissions. This adds to the cost base of those who need more credits, making their carbon emitting behaviour more expensive, while those who are able to sell extra credits are rewarded for having cut their emissions.

The system of traded credits should provide an incentive for companies to cut their emissions, to cut their cost base and generate revenue from the sale of extra credits. It was also intended to provide a means of transfering technology and funds to the developing world through investment in offsets. Using such a market to establish proper incentives to improve global carbon emissions levels could benefit both developed and developing economies, thus contributing to a reduction of global emissions at a lower total cost.

The original idea was that the cost of implementing carbon saving measures would be offset by funds generated from the later sale of credits. This has become a complex area, as there are regulated and voluntary markets, and a plethora of different credits available for sale. At the same time, early attempts to set up such markets have been clearly flawed and we'll look at what we've learned in how to better structure such a market.

The concept of the carbon market is enshrined within the Kyoto Protocol, the current international agreement on addressing the causes of climate change. We'll take a look at the framework of the agreement, its strengths and weaknesses, as well as at what needs to be agreed for an international treaty to succeed it in 2012. While a successor to the current Kyoto Protocol is a vital part of the process in bringing people together to address climate change, there is also an argument for unilateral and individual action. We'll explore how emissions reductions are being tackled outside the framework of a global treaty, and what we as consumers, corporations, communities and countries can really do to make a difference.

Many believe that cutting carbon from our economy will require a move to an earlier level of subsistence living, or at the very least the creation of new technological approaches to managing the problem. The reality is that we can transform our consumption patterns with technologies available today, and still maintain a level of comfort that should more than suffice to support a decent quality of life. We'll also look at the technological and cultural transitions that we may need to make if we are to achieve a transformation to a low carbon economy.

The carbon markets offer us an opportunity to transform our economic framework and bring a connection between rights and responsibilities back into the marketplace. We have the means to change the way the world works, if we understand the stakes, implement the mechanisms correctly and take the decision to act.

Climate Change and Greenhouse Gases

1

Most people would agree that greenhouse gas (GHG) emissions are increasing, with the impact felt from the Artic to the Amazon. But for many of us, we're not exactly clear about what that means, or how it is likely to affect us.

The Intergovernmental Panel on Climate Change (IPCC) is a panel of over 2,000 scientists set up to provide information on global warming and advise the international community on the best approach to addressing climate change. Increasingly the IPCC, whose job it is to analyse peer-reviewed research in the world's scientific journals, has reported that evidence suggests that global temperatures are on the increase, driven by increasing amounts of GHG in the atmosphere. If we do not act to cut the emissions of GHG, we may be responsible for rapid changes in the global climate.

The last report from the IPCC to provide an integrated view of climate change, the *Fourth Assessment Report* (AR4) published in 2007, said that warming of the climate system is now unequivocal and warned that 'changes in atmospheric concentration of greenhouse gases (GHGs) and aerosols, land cover and solar radiation alter the balance of the climate system'. The argument is that as the levels of greenhouse gas continue to rise, so does the temperature. That doesn't mean that it is always going to get warmer – what it means is that the climate is going to change more rapidly. The higher the temperature increase, the more dramatic changes the climate is likely to experience.

The greenhouse effect itself is a natural process essential for life on earth. It ensures that the planet is warm enough on which to live and that the atmosphere provides sufficient protection from the Sun. However, manmade emissions of GHGs are enhancing its effect, speeding up the warming of the world. Climate change will have an impact on air quality, biodiversity, forestry, water potability and supply, increased desertification and more.

In climate change terms, carbon dioxide (CO_2) is considered the most important GHG, as it is the most common. In common parlance, 'carbon' is the term used to describe the group of GHGs, which include methane (CH_4), nitrous oxide (N_2O), halogenated fluorocarbons (HCFCs), ozone (O_3), perfluorinated carbons (PFCs), and hydrofluorocarbons (HFCs), and are described in terms of their equivalent to carbon dioxide – CO_2e.

The concentration of CO_2, the most prevalent GHG, in the atmosphere today is at the highest it has been for the past 650,000 years. Levels of CO_2 in the atmosphere are understood to have been around 280 parts per million* (ppm) prior to the Industrial Revolution (c.1760–1860) in Europe. Levels of CO_2 in the atmostphere in 2008 were reported to be over 385ppm and, if you include all GHGs covered under the Kyoto Protocol, levels of CO_2e are already at 440ppm and understood to be rising at around 2ppm per year.

Put another way, overall figures from the US Energy Information Administration suggest that while some headway has been made in cutting emissions from developed countries, global emissions are currently rising at about 3% a year. The US Department of Energy (DoE) has estimated that global CO_2e emissions in 2010 will in fact have risen over 40% against 1990 levels.

In just 250 years we've released more than 1,800 billion tonnes of CO_2 into the atmosphere – generated mostly through the burning of fossil fuels, with most of the rest generated from changes in land use (mostly deforestation), in building cities and developing farmland.

---

* Parts per million (ppm) simply means the concentration by volume of any given substance in other (carbon dioxide in the atmosphere in this case) and is the way that pollution levels are normally described.

It's been estimated that about 1,000 tonnes of CO_2 are released into the Earth's atmosphere every second. And these gases will stay around in the atmosphere for years, so any cuts that we make will take years to make a significant difference – about 100 years in the case of CO_2.

Plants, soils and oceans currently absorb about half of the CO_2 emitted by human activities, limiting rises in atmospheric CO_2 and slowing global warming. As temperatures increase, this absorption is very likely to

'In just 250 years we've released more than 1,800 billion tonnes of CO_2 into the atmosphere'

decrease. For example, plant matter in the soil often breaks down more quickly at higher temperatures, releasing carbon more quickly, and amplifying any warming trends.

According to the United Nations Environment Programme (UNEP) 2009 Yearbook, the Greenland ice sheet, which could raise sea levels by 6 metres if it melted away, is currently losing more than 100 cubic km a year, a rate faster than can be explained by natural melting. The IPCC 4th Report argues that if we can limit growth in CO_2e to 450ppm by 2050 we have a 50% chance of limiting global average temperature rise to a 2°C (35.6°F) by 2100: levels of 550ppm of CO_2e in the atmosphere could result in increases of 3–4°C (37.4–39.2°F) degrees. Some scientists disagree, with many claiming that the IPCC figures are conservative.

According to data presented in Copenhagen at the March 2009 International Scientific Conference on Climate Change (ISCCC), the climate system already appears to be moving beyond patterns of natural variability. These parameters include global mean surface temperature, a rise in sea levels, ocean and ice sheet dynamics, ocean acidification and extreme climatic events. There is a significant risk that many of the trends will accelerate, leading to an increasing risk of abrupt or irreversible climatic shifts.

A group of scientists led by Dr James Hansen, a director of the NASA Goddard Institute for Space Studies, claims that in order to avoid catastrophic climate change we need to cut CO_2e levels to

350ppm by 2050, not 450ppm. Scientists may disagree about the specific details and likely consequences of climate change, about the efficacy of climate models and the likely environmental response to an increase in atmospheric temperature but the majority today agree that to lower the risk of global warming, we should cut our emissions.

Today roughly 50 billion tonnes of CO_2e is released into the atmosphere every year and it is the scale of these emissions that must be addressed. The IPCC has said that these levels must fall to around 20 billion tonnes per year by 2050 if there is to be any real chance of keeping the global temperature increase down to 2°C (36°F). That means that if we are to stabilize our emissions, and ultimately maintain climate balance, we need to bring annual global emissions down to more than 80% below current levels.

This is no easy feat and, to put it in perspective, if today's emissions were divided by the population (at around 6 billion), the average emissions per person would be roughly 8 tonnes per year. With the global population expected to have grown to 9–10 billion by 2050, this would mean each person alive in 2050 could only emit 2 tonnes per person to achieve the IPCC goal. This is going to mean big changes, in rich, poor and middle income countries alike.

The problem with climate change

Of course, climate change itself is not actually the problem – the climate is an active system and it continues to change over time. The impact of rising temperatures could see the glaciers of the Himalayas, Andes, Rockies and Alps begin to melt, affecting the flow of clean water to land all over the world. It's the potential for climate instability, for which we're unprepared, that is of most concern.

Seasonal changes in temperature can also affect the ability of crops to grow, thereby our ability to feed ourselves and having a direct impact on the ability of a range of species to survive. The increasing acidity of the oceans (as more CO_2 dissolves in the sea-water) could influence water eco-systems and their ability to support sea-borne life. At the same time, increasing volatility in weather systems could result in ever more violent cyclones and hurricanes – there is precedent for this: a storm surge in Bangladesh killed 300,000 people in 1970 and further surges killed 200,000 people there in the 1980s.

The range of temperature increase that the IPCC predicts could result from anthropogenic emissions is between 1.1 and 6.4°C (33.9–43.5°F). The last time global temperatures were 5°C (41°F) above where we are now was the Eocene period, around 50 million years ago, when the world was mostly swampy forest, and there were alligators at the North Pole. The last time we were 5°C below where we are now was very recent, about 10,000 years ago (the last Ice Age) and as we warmed up from there, the UK separated from the Continent and most of the world's rivers were redrawn.

Globally, even a 4°C (39.2°F) temperature rise could have a cataclysmic impact. The 2006 *Stern Review on the Economics of Climate Change*, compiled by Nicholas Stern, a former World Bank economist, predicted that increases of that level would see between 7 and 300 million (dependent on the increase in sea levels) more people affected by coastal flooding each year, a 30–50% reduction in water availability in southern Africa and in the Mediterranean, a fall in agricultural yields of between 15–35% in Africa and that 20–50% of animal and plant species would face extinction. Even a 2°C (35.6°F) increase, which some now believe to be inevitable, could have a significant impact on the ocean's ability to absorb CO_2 from the atmosphere.

As the IPCC gears up for the publication of the *Fifth Assessment Report* in 2014, the scientific community is reviewing the latest data. Reports presented at the ISCCC in March 2009 suggest that temperature increases of around 5°C (41°F) are already far more likely than previously believed. Professor John Schellnuber, of Germany's Potsdam Institute for Climate Impact Research, warned that even if this was limited to 2°C (35.6°F), the climate impacts could be more dramatic. He stated that emissions are growing faster than anticipated, carbon sinks (forests, atmosphere and oceans) are storing carbon less effectively, and that 80% cuts in emissions by 2050 are now critical if we are to avoid a 5°C temperature increase in the longer term.

If emissions don't start to decrease soon, we run the risk of fundamentally changing the underlying balance of the climate. The problem is the rate at which climate change could accelerate and the social, environmental and economic consequences that this would bring. It is the potential impact of changing weather patterns, sea level rise and extreme weather events on our ability to inhabit certain areas, provide sufficient food and clean water to the global population and find sufficient sources of fuel to power our economies. The problem is the impact of climate change on the already strained confluence of the world's finite resources, increasing pollution and population.

Our dilemma lies with the economic impact of such changes on individuals, corporations, communities and nations. With resources becoming more scarce, geopolitical tension has increased. As a result, we need to find new ways of managing those resources, an approach which is equitable, economic and efficient – and likely to be agreed upon at an international level.

Why conquering carbon is critical

Cutting our emissions may be a vital step in the evolution of our society, as it is a key step in taking responsibility for our own actions. How we choose to address the problem could be the most important thing. We need to find ways of driving changes in behaviour, to agree on a common framework of values that benefit the welfare of the world as a whole. If this is done right, we could learn to ascribe value to assets outside the traditional economic framework – those things which support life on earth and are therefore the underpinnings of any economic systems – clean air, water, food and energy.

Climate change is a global problem that requires a global solution. It doesn't matter where CO_2 is emitted, GHGs have equal effects on climate change irrespective of where they are emitted – it ends up in the atmosphere that we all share. The world is interconnected – we have to look at what's going on around us because every action is now about the impact we have on each other.

We need to move quickly if we are to have any chance of significantly altering the trajectory of economic development. If we are to achieve a goal of a low carbon society which retains quality of living for all, we need to act now.

We know that difficult choices will have to be made if we are to stop the irreversible and dangerous impacts of climate change. It will take profound changes in human behaviour to bring about improvement in every area where the global environment is threatened. The challenge lies in how to affect a rapid transition in the economy while maintaining our quality of life. The Board of the Swiss Federal Institute of Technology in Zurich launched the 2,000-watt Society in 1998. It offers a vision in which every person in the developed world would cut their overall energy consumption to 17,500 kw hours, today's worldwide average. The level of 2,000 watts is the current global average of power consumption per capita, although local averages are around 6,000 watts in Europe, while some Asian and African countries consume only a fraction of this amount.

This level of consumption is equivalent to a continous consumption of 2,000 watts, spread over living and office space, food, power, travel and public infrastructure. This would require a reduction in energy demand per capita by two thirds within industrial nations. The scientists responsible for the research said they believed the goal to be 'technically feasible', especially given that today around two thirds of primary energy demand is lost in energy conversion to power appliances.

Researchers believe that these cuts to global average power consumption can be achieved without massive changes to standard of living, despite a projected 65% increase in economic growth by 2050. That means that a level of existence including cars, plane travel, television and the majority of modern conveniences could be available to all, through a transformation of our consumption patterns and energy environment.

Such cuts in power demand could be achieved through a higher degree of material and energy efficiency, new corporate concepts, the wide-spread implementation of renewable energy, the refurbishment of buildings and equipment, vehicles and other facilities, combined with a more efficient use of resources.

The first step in attempting to address this issue has already been taken, with an international agreement (the Kyoto Protocol) put in place to encourage countries to cut their emissions. Some argue that it is too little, too late. Many major emitters, such as the US and Chinese economies, have no comittment to cutting emissions – the US because it refused to sign the Kyoto Protocol and China because it was given no caps on its emissions. The current negotiations on an agreement to follow Kyoto, which expires in 2012, are critical for setting the framework in which the global economy moves forward. One key element of this negotiation is working out how best to combine pricing carbon out of the economy while maintaining economic growth, and fairly sharing the burden of the transition away from carbon-intensive development.

If the mechanism is successful – if pricing carbon changes the way we use it – we can use the same model as a wider resource management tool. If we can put a price on our limited resources, find a way to allocate their usage fairly and put a steep price on any excess use, we might be able to make a real difference.

The cost of climate change

Carbon and the cost of climate change has become a mainstream issue as a result of the economic analysis of the cost of climate change. Environmentalism has been growing in strength over the last 40 years but it is only when the potential cost becomes clear that governments have been galvanized into action.

The *Stern Review on the Economics of Climate Change*, compiled for the UK Treasury in 2006, is one of the most influential political documents with regard to climate change. Nicolas Stern's 2006 analysis suggests that if we carry on under a 'business as usual' scenario, there is a 50% chance that temperatures will rise over 4°C (39.2°F) over the next 100 years. According to the Review, if we stabilize emissions at 500ppm CO2e, that 50% probability drops to about 3%. Addressing climate change then becomes a question of risk management.

The cost of action will be high but, according to Stern's economic analysis, not nearly as high as doing nothing. He suggests that as little as 1 to 2% of global GDP (the total market value of goods and services, also known as gross domestic product) invested annually could help solve the problem and economics on the ground. That's a cost of around $500 billion to $1 trillion if we take action today. In 2008, economist Ross Garnaut was commissioned to analyse the cost of climate change on behalf of the Australian government; and he worked out that the cost of cutting carbon in Australia would be less than an Australian dollar a day if action is taken now. According to Stern however, if we fail to take action promptly the cost could be as high as 20% of global GDP by 2050.

'Nicholas Stern suggests that if we carry on under a "business as usual" scenario, there is a 50% chance that temperatures will rise over 4°C (39.2°F) over the next 100 years.'

At the March 2009 ISCCC, Stern warned that the cost of inaction could be 50% higher than his earlier predictions, equivalent to around one-third of global wealth.

Part of the problem is that we don't really know what the cost of carbon is going to be. In order to set a real cost for the emission of a tonne of GHG, we would need to know the long-term cost of the impacts of climate change, which are still being estimated. The work of economists such as Nicholas Stern, Ross Garnaut, Samuel Fankhauser, Robert Stavins, William Nordhaus and others on the impact of climate change, has enabled us to quantify the impact of GHG emissions as an externality (meaning the future impact on others of certain actions, or the costs of creating a product or service which are not included in the price today). These can then be factored into a cost benefit analysis of changing behaviour to mitigate climate change.

The cost of these externalities are the financial cost of mitigating and adapting to climate change, and increased costs associated with access to resources and commodities. These can range from increased input costs for energy fuel sources, to the costs of the construction of dikes and seawalls to protect against rising sea levels and extreme weather events. Changes in patterns of droughts and floods, heat waves and frosts could destroy forests and agricultural systems and sudden temperature shifts could, at the very least, make buildings without the requisite insulation completely uninhabitable, changing even the areas of the world in which people can live.

Policymakers have predominantly focused on the mitigation of climate change in the last few years, the idea that by developing means of cutting GHG emissions, we can cut the likelihood of dramatic climate change. This has included the introduction of emissions caps, energy efficiency and a focus on accelerating the use of renewable energy, etc. It is becoming increasingly clear, however, that the cost of adapting to climate change is going to need to be clearly factored into any economic equations. This could range from something as simple as building sea defences around low-lying land, to the development of new sustainable forms of agriculture. The cost of supporting change in the world's poorer economies, who might not be able to afford the measures required, also needs to be considered. All these changes will have to be paid for one way or another.

The question is who will bear the cost? Many critics of environmental action complain that the costs will constrain economic growth. Yet if the early 21st-century credit crunch has shown us anything, it should be that constant and increasing borrowing eventually grinds the entire system to a halt. If the way we are treating the global resources of the Earth consists of nothing more than constant and increasing borrowing from the future, we can see that we're setting ourselves up for a fall.

Although we have the technologies in place to transform our economy and our approach to resource consumption, for many people it appears more expensive and time consuming to seek out low carbon alternatives than it is to carry on as usual.

For most of us, especially in an economic downturn, the concern is obviously about how much it is actually going to cost us. Is power going to be too expensive? How will I buy food? Will the things I need to buy go up in price? It's unlikely that the cost of addressing the problem will be spread fairly between individuals, so we need to understand whether or not we can afford not to spend the money now. While the street philosopher might believe that the greatest good for the many outweighs the needs or wants of the few on a society-wide basis, change sometimes needs to be encouraged.

Why do we need a carbon price?

One of the most effective ways of changing people's behaviour in a capitalist economy is in the pricing of goods and services. By including an explicit price of carbon we hope to discourage and curb its emission. At the same time, in a predominantly capitalist world, the use of market forces to effect modular change has a great deal of appeal – it suggests that economic constraints in one area can be offset by investment opportunities in another.

Price plays a critical role in encouraging change in a market economy and we have to use all means at our disposal if we are going to make any headway. Price gives investors and corporations a signal, enabling them to make longer term decisions about investment in research and development (R&D) and operational streamlining. It encourages technological change, because as new technologies are invented and deployed, costs fall as the scale of deployment increases, making them more affordable.

The idea behind a price for carbon is that if we charge emitters of CO_2e money for those emissions, then they will be encouraged to either cut their emissions, or find a new way of doing business. That is the intention behind including the cost of externalities in the price of products and services to the end user. Certainly the current prices of gasoline, electricity and fuel, in general, include none of the costs or externalities associated with climate change and if they did, their prices would be significantly affected.

There are many difficulties in setting a price for carbon, not least of which is the fact that various economists have different arguments as to the cost of climate change. As a rule, the idea is to look at what it might cost in the future to adapt to a new climate, to new resource restrictions, and then discount that cost back down to the present day. The biggest problem, however, is the idea that a cost tomorrow is less important than a cost today. In October 2008, the Bank of England's *Financial Stability Report* cited that the credit crunch will cost the taxpayer £1.8 trillion and governments around the world are attempting to address the problem with economic stimulus packages, tax breaks and other fiscal policies. But apathy and lack of action in fighting climate change could cost the global economy $20 trillion dollars (£11 trillion) a year by the end of the century. One of the problems is that the cost of action on climate change is immediate, whereas the cost of inaction will only have an effect in the future.

It is widely accepted that in order to effect significant changes in behaviour, the cost of carbon must be explicit and long-term, while, at the same time, sufficiently high as to make it more cost-effective to cut emissions than not. Many of the more popular forms of industrial carbon abatement (for example carbon capture and storage) are not considered to be commercially viable at a cost of less than €32 (US$40) per tonne of carbon. When the carbon price hit its peak in mid 2008 there were expectations that the cost of emitting CO_2 would have a significant impact on changing patterns of emissions. The crisis in the financial markets has led to an economic slowdown, however, with declining industrial production and electricity demand leading to lower demand for emissions credits. This has seen the price of credits fall dramatically.

It's vital that we have a price for carbon to stimulate action sooner rather than later. The sooner action on emissions is taken, the more likely we are to avert the worst effects of climate change, but it is increasingly difficult to justify the expenditure on new technologies and renewable alternatives given that we seem to be entering a period of economic depression. Can we afford to invest copious amounts of money in mitigating climate change today when financial systems around the world are collapsing? The answer is simple, and lies in the question. It is not what investment we can afford today, but whether we will be able to afford the consequences tomorrow.

Setting a price on carbon is unlikely to work isolation. Simply assuming that if we increase the cost of a product people won't use it, ignores basic human nature – we tend to stick to what we know, and we don't respond well to change. We also need to use regulatory and legislative frameworks that encourage positive changes in behaviour. We need to find the right combination of price, regulation and philosophical approach to ensure that the goal of cutting CO_2 emissions is achieved.

Equitable change

One of the most contentious issues is how to transform the overall worldwide economy to a low carbon basis in a way that is fair. The problem of climate change is a global one, as pollution, land degradation, deforestation and floods have an impact far beyond their immediate environs. What seems like an isolated event is always linked to another environmental process, and part of a greater chain. This is one reason why international agreements and policies are so important.

The consensus among experts is that the best way to address climate change is to keep CO_2 levels down, but the reality is in the current global economy, that's going to be hard, as industrialization and modernization in the last 200 years have been based on carbon-intensive industries such as oil and coal. There are some easy solutions to the problem on offer if, for example, you're happy about reducing the global population by 80%, but on a human basis, that suggestion isn't particularly reasonable, or even practical. The question is whether or not people have got enough common sense to put long-term benefit ahead of short-term desires.

Another key issue is ownership of the finite resources of fuel, fertile land and clean water – who do they belong to and what rights do these owners have? The questions we must ask are whether these rights belong to the government, the landowner or the indigenous people that were displaced for their use? Do the rights to resources belong to those who exploit them, or those who can afford to pay for them? Do they primarily belong to this generation or the next? If we decide that coming generations have the right to access resources after we're gone, then the cost of safeguarding those resources must be factored into the cost of emitting GHGs.

We know that the effects of climate change are likely to hit hardest in the developing world, an area least able to afford the costs of mitigation or otherwise protect themselves against climate-related damages, where income is low. This is especially true in developing economies with relatively large agricultural sectors in flood plains like Bangladesh or drought-prone regions such as Darfur. Many in the developed world demand the larger developing world economies limit GHG emissions in a reasonably equivalent way. Yet, given that 80% of man-made greenhouse gas is believed to have been generated by the developed world in the last couple of hundred years, doesn't it seem a little unfair to keep a large proportion of the world's population in poverty because we don't want to turn the heating down?

The developed world geared up its economies through the consumption of cheap, carbon intensive fuel and countries such as China and India understandably argue that they should be allowed the same development opportunities. Yet, now we've got to cut the emissions associated with that economic framework – do we do this unilaterally or must it be a global agreement? Many still argue that there's no point in bothering to take action on GHG emissions without developing economy emissions limits – after all, any changes in our carbon emissions will be dwarfed by the impact of population increases and economic growth in China, defeating any positive benefit we could create.

While calculations can differ, according to the Netherlands Environmental Assessment Agency in 2007, China's CO_2 emissions contributed the bulk of a 3.1% global rise, with an 8% national increase. In a statement the Agency said 'With this, China tops the list of CO_2 emitting countries, having about a quarter share in global CO_2 emissions (24%).' The US was second with 21%, while the European Union come in at 12%, India 8% and Russia at 6%.

What seems to be ignored here is whether or not it is reasonable to expect people to continue to live in poverty because we have recently discovered that economic growth has been found to have unforeseen and unpleasant consequences. More importantly, research is increasingly showing how much of China's emissions, for example, are associated with manufacturing goods for export. Research shows around one-third of China's emissions are related to consumption in the developed world. That doesn't mean that increasing population growth and consumption in the developing economies is a good thing, but it does mean that we need a greater understanding of the emissions cycle before we start to ascribe responsibility and blame.

Some people believe it will be impossible to change the world's current dependence on fossil fuel, at least in any meaningful timeframe. But if we are serious about transforming our economy, we need to find new ways to address the needs of our societies, new ways to encourage beneficial changes in political, economic and personal behaviour.

There is no question that the growth in emissions from the developing world must be addressed but there are already processes in place to tackle this. One of the key mechanisms of the Kyoto Protocol is intended to promote the transfer of technologies to developing markets so that their economies can become more efficient as they grow.

At an international level, countries are coming together, trying to agree reasonable and fair international agreements to mitigate carbon emissions. Individual countries, companies, communities and consumers are beginning to take action on their own. If we work together, using the tools that have worked in the past, with good will and a fair assignment of value, we could transform the world.

An international emissions reduction agreement was agreed in 1994 – the Kyoto Protocol. Ratified by 183 countries, it set targets for the developed world to reduce their emissions by 5% below 1990 levels by the 2008–12 period. More than that, it enshrined the concept of using the carbon markets as a means to cut global emissions. The carbon markets are the first attempt by the international community to use pricing and trading as a way of trying to change the odds of significant climate change.

Agreement of the Kyoto Protocol was a huge step forward in addressing the issues of climate change mitigation but progress on a post-Kyoto agreement has been slow, and fraught with difficulty. The issues at stake are too great to let responsibility lie solely with negotiators of an international agreement that must take into consideration the needs of different national economies, their fears and their pride.

Actions taken at every level of society have a critical role to play. On an individual basis, the idea that not using a plastic bag is going to stop the polar ice-caps melting may be a bit difficult to swallow, but the fact remains that the actions of millions of individuals together can make a huge difference. It is not a question of one small action having no impact, but an acceptance of the fact that with carbon emissions, everything is connected. It doesn't matter where the gas comes from, it is here to stay for a long time. That means that every small action does make a difference – the key is to encourage billions of those small actions, and millions of larger ones.

On a corporate level, if we, as consumers, can encourage companies to make sustainability and carbon management key aspects of corporate value, then we will transform the way in which our economy works. As communities and cities we can find ways to transform our expectations and manage our resources closer to home. As countries, we need to focus on environmental solutions for climate-change problems which can support economic growth and energy independence. If we work together at every level of society – national governments, companies, communities and individuals – then we have a great potential for change.

The mechanisms that we have in place are about encouraging such actions across all levels of society, because they are about the transformation of how we assign value. We are being offered an unique opportunity to move away from the selfish consumption-driven view of modern society, towards a place where value is assigned for its purpose within the whole.

Transition to a low carbon economy

Wherever you sit in terms of the climate change discussion, one thing is irrefutable – the levels of GHGs in the atmosphere are on the rise. If you accept the scientific consensus of the IPCC, we run the risk of dramatic climate change if we don't act soon to stabilize, and then cut, carbon emissions.

For many the idea of needing to cut carbon emissions so dramatically makes it seem like a futile task and there are a plethora of unanswered questions to follow this undertaking, for example: how are we supposed to cut emissions? Can we cut them in time? Will cutting emissions enough to make a difference mean fundamental changes to our way of life? What about mitigation – can we alleviate the effect of climate change? If we accept that there is a problem, what exactly are we supposed to do about it?

The list goes on. Why do we need to understand what difference it makes that a group of international policymakers signed a treaty and agreed to treat an invisible odourless gas as a commodity? What are our governments doing with our money and why? When companies try to pitch their wares to us on the grounds that they are greener – what does that actually mean and should we be paying for it? Can we, as individuals, make a difference and if so, how? These are issues which we will explore later in more detail.

To implement the necessary changes and transform our economy to a low-carbon, sustainable basis, we need to find a way of thinking in the long term, no longer subject to the short-term demands of political appointment or shareholder expectation. That's going to require new forms of international co-operation, and the development of ways to ensure good governance and the implementation of agreements.

We need to deploy technologies to cut carbon or increase carbon productivity and invest hundreds of billions in deployment, in such a way that we support the economic and political stability of individual nations. We need to find ways of defending and developing our forests and sources of biodiversity. On a personal level, we need to make fundamental changes in lifestyle and habitation, both in terms of style, place and expectations of comfort. Most of all, we need to find a way of doing all these things that is effective, efficient and equitable.

This might seem like an overwhelming task but changes are already beginning to happen and what we need to do is understand what technological tools and policy frameworks we already have to offer, and how to extend that experience. What we need to see is a transformation of business practice, personal expectations and a growth in responsibility. We have the knowledge as individuals and communities about what needs to be changed and the technology to effect that change. Changing people's behaviour is perhaps the hardest thing to do and we will look at the options that have been explored at a national and international level. It is as individuals, and groups of individuals, that we can have the most effect. Whether as voters, consumers, employers, shareholders or politicians, we will explore the options that countries, cities, companies and consumers are trying and look at what we can do ourselves.

Firstly, we need to gain a better understanding of just what the long-term environmental and economic impact of climate change might be.

The Resource Wars

2

While there are those who believe that large proportions of the human race can be moved to act altruistically for the 'greater good', it is a truism that for many altruism is a luxury. For the larger part of humanity, it is perhaps more fair to say that we need to understand why we need to change how we behave, before we consider whether or not we are prepared to do so.

The environmental impacts of our addiction to fossil fuels are well known – rising GHG emissions that are likely to lead to an increasingly hotter world with extreme weather conditions. The problem lies in understanding what the impact of rising sea levels in the Indian Ocean or the increasing desertification of the Sahel region of Africa will mean. Environmental changes can have real implications for poverty reduction, economic development and national security. They can result in increased economic migration, diminished access to energy, as well as failing supplies of clean water and food.

One example is the genocide in Sudan, where climate change is believed to have played a role. Access to water supplies is understood to have provided a spark for the outbreak of civil war in 2003, as droughts affected grazing lands and people moved from their traditional homes, coming into conflict with those whose lands they entered. The tragedy, which has seen killing on a vast scale, has also resulted in the destruction of foodstocks, seeds, livestock, wells and irrigation systems, making the region uninhabitable, and with refugees and displaced persons running into the millions.

As clean water, food and fuel become harder to access, this will become a critical issue for all nations. Whether personally affected or not, growing unrest could disrupt global supply chains, and international aid will be strained in attempting to mitigate the worst effects of displacement. Imbalance in the food supply is already having an effect. According to the UNEP in 2007, the upsurge in the prices of food grains cost developing countries US$324 billion, the equivalent of three years worth of global aid.

The human and financial cost of climate change, as well as access to valuable resources, are likely to become key variables in civil wars and national aggression during the coming century. By making explicit the human and financial costs of inaction on climate change we can make clear the consequences of our choices and hope to stimulate action while there is still time

Even discounting the added impact of climate change, global resources are already under pressure, whether in terms of oil supply, food or water. Competition over energy sources is already reshaping the geopolitical map – supplies are being used as political bargaining tools, while the enormous growth in oil and gas revenues in the last couple of years has given increasing political strength to energy-rich states, some of which are politically unstable. In fact, there are rising tensions over commodities and resources everywhere.

Living beyond our means

According to the WWF's 2008 Living Planet Report, over three-quarters of the world's population live in such a way that their levels of consumption outstrip the ability of the local environment to renew itself. According to the report's ecological footprint analysis produced by the Global Footprint Network (GFN), global biocapacity (meaning the amount of land and sea needed to produce our resources, absorb our waste and capture our emissions), is an average of 2.1 hectares (ha) per person. Yet the current average global footprint per person is already at 2.7ha. This led James Leape, the WWF International director-general, to conclude in 2008: 'If our demands on the planet continue to increase at this rate, by the mid-2030s we would need the equivalent of two planets to maintain our lifestyles'.

The report highlighted that in 2007, the US and China had the largest national footprints, each consuming in total about 21% of global biocapacity, but US citizens each required an average of 9.4 ha (or nearly 4.5 Planet Earths if the global population had US consumption patterns) while Chinese citizens use on average 2.1ha per person (1 Planet Earth). The UK was somewhere in the middle, with per person consumption at around 5.3ha – still over twice the 2.1ha per capita sustainably available to the global population.

Part of the problem is that biocapacity is unevenly distributed: eight nations, the US, Brazil, Russia, China, India, Canada, Argentina and Australia, contain more than half the world's biocapacity. A combination of population and consumption patterns make three of these countries ecological debtors, with footprints greater than their national biocapacity – the US (footprint 1.8 times national biocapacity), China (2.3 times) and India (2.2 times). Compare these to the Congo, which has the seventh highest per person biocapacity of 13.9ha per person and an average footprint of just 0.5ha per person, and we can see a future of degrading biocapacity from deforestation, while demand for increasing biocapacity is driven by a combination of rising population and export pressures.

Another area covered in the report which indicates the scale of the problem is the Living Planet Index (LPI). Compiled by the Zoological Society of London (ZSL), the LPI shows a nearly 30% decline since 1970 in nearly 5,000 measured populations of 1,686 species. These dramatic losses in natural wealth are being driven by deforestation and land conversion in the tropics (50% decline in tropical LPI over the period), as well as the impact of dams, water diversions and climate change on freshwater species (showing a 35% decline). Pollution, over-fishing and destructive fishing in marine and coastal environments are also taking a considerable toll. While the cynic might question why we should care about species dying out, dramatic changes in species variation can have an equally dramatic, and often unexpected, impact on the biosphere, ranging from soil quality to maintenance of the food chain.

Carbon emissions from fossil-fuel use and land-use change are the greatest component of humanity's footprint, once again underscoring the threat of climate change. The WWF report recommends a re-think on natural resources. 'Decisions in each sector ... must be taken with an eye to broader ecological consequences. It also means that we must find ways to manage across our own boundaries – across property lines and political borders – to take care of the ecosystem as a whole'.

'We are acting ecologically in the same way as financial institutions have been behaving economically – seeking immediate gratification without due regard for the consequences,' said ZSL co-editor Jonathan Loh. 'The consequences of a global ecological crisis are even graver than the current economic meltdown'. The report says the single most important solution to the 'eco crunch' is CO_2-emission reductions by up to 80% by 2050.

This is clearly no longer purely an environmentalist's dream. The intention to reduce emissions by 80% from 1990 levels was set as a target in the UK's 2008 Climate Change Bill and President Barack Obama has said the US should also aim to reach that goal. There is a growing acceptance of the correlation between protecting resources and national security.

Extreme weather events and social disruption

Extreme weather events are, in the public mind, a clear indicator of the impact of climate change. Given the complicated nature of both the global and local climate systems, it can be difficult to know whether extreme weather is being caused by changes in short-term trends or long term climate change. Many extreme weather events are natural variations in weather patterns, which can occur in long-, medium- or short-term cycles. It's possible that there is no link to climate change, no matter how logical that connection might appear.

No matter the truth of the complexity of climate science, extreme weather events provide a visceral prism through which we can view the potential consequences of dramatic climate change. Most importantly, they illustrate the dangers in just how unprepared we are to address dramatic climate events.

In 2003, much of Europe was hit by unprecented heatwaves. France, where over 14,000 people (predominantly elderly) died from heat related causes, was heavily affected, and the heat waves became a cause celebre in the failure of the French authorities to react promptly during August. Temperatures of over 40°C (104°F) were recorded. Many buildings were not fitted with air conditioning and people were unprepared for managing high temperatures, with many failing to hydrate properly. Reports claim that over 50,000 people died from heat related deaths across Europe. Yet current IPCC models predict that on current trajectories it is possible that the average summer temperature in Europe will reach similar levels by 2040. While it may be possible to plan for such eventualities through increased use of energy-intensive air-conditioning programmes, this will do little to protect agriculture or local ecosystems.

At the same time, few will have forgotten the 2005 devastation of New Orleans following Hurricane Katrina – the loss of life and property damage from which the city has still not fully recovered. The failure of the US administration to address the problems after the fact highlighted for many the devastation created by the aftermath of a disaster, rather than simply the direct impact of a weather event. The event caused the temporary displacement of 1.5 million people: around 300,000 have never returned home. The disaster made clear that in the long term, it is going to be the ability of countries to respond to such crises that will define their impact on citizens.

It will be our failure to address the impact of such climate events, our inability to deal with the consequences that could wreak havoc – environmentally, socially and financially. As such climate related events increase in frequency, so will their impact on an already strained international aid infrastructure, and on the need for individuals to move to safer ground. This threatens much more than climate and energy security – it endangers geopolitical stability.

Conflict and climate change

The UN Security Council held its first debate on climate change and conflict in April 2007. The session was chaired by then British Foreign Secretary Margaret Beckett, who posed the question: 'What makes wars start?' Given that recent scientific evidence had reinforced some of the worst fears about climate change, she warned of migration on an unprecedented scale because of flooding, disease and famine. She cautioned that drought and crop failure could cause competition for food, water and energy and that climate change is about 'our collective security in a fragile and increasingly interdependent world'.

At the same time, UN Secretary-General Ban Ki-moon outlined several 'alarming but not alarmist' scenarios to do with climate change. He said, 'Throughout human history, people and countries have fought over natural resources. From livestock, watering holes and fertile land, to trade routes, fish stocks and spices, sugar, oil, gold and other precious commodities, war has too often been the means to secure possession of scarce resources. Even today, the uninterrupted supply of fuel and minerals is a key element of geopolitical considerations.' The dangers include limited or threatened access to energy increasing the risk of conflict, a scarcity of food and water transforming peaceful competition into violence, and floods and droughts sparking human migrations, polarizing societies and weakening the ability of countries to resolve conflicts peacefully.

Some UN representatives, including those from China and Pakistan, argued that the Security Council was not the proper forum for a climate change debate, as it was an issue of sustainable development. However, Papua New Guinea's representative, speaking on behalf of the Pacific Islands Forum, said that as a result of climate change Pacific island countries were likely to face massive dislocations of people, similar to the population flows sparked by conflict. He warned that the impact on identity and social cohesion was likely to cause as much resentment, hatred and alienation as any refugee crisis.

The impact of environmental change change is a growing concern amongst those non-governmental organizations (NGOs) and international aid bodies which deal with the long term impact of population displacement. One in ten people – one in eight of those who live in cities – live 10 metres or less above sea level, and near the coast. That means 600 million people living in low-lying coastal zones, the 'at-risk' zone for coastal flooding. The IPCC said, in 2007, that sea levels were likely to rise by nearly 60cm by the end of the century. However, according to research done at the Potsdamn Institute for Climate Impact Research in Germany in 2007, the IPCC figures were conservative, compared against actual subsequent changes.

Many of the poorest areas of the world are those at risk. The Potsdamn study found that around 75% of those living in the 'at-risk' zones are in Asia, 21 nations have over half their population in these regions, and that nearly two-thirds of urban areas with over 5 million inhabitants are at least partially at risk. The global average of cities on the coast is 7% but rises to 12% in Africa. The 10 countries with the largest number of people living in this vulnerable, low-elevation zone, include in descending order: China, India, Bangladesh, Vietnam, Indonesia, Japan, Egypt, the US, Thailand and the Philippines.

The danger of rising seas and the potential for increased salination of coastal acquifers and surface water is a concern all over the world. A prime example is Bangladesh, which is famously affected by storm surges. Many of the farms based in its silty coastal areas have poor irrigation practices and this, combined with the impacts of deforestation, has resulted in increasing saltiness of the water in coastal areas. This has a direct impact on the local ability to grow food. It is said that if the sea rises 1 metre, 20 million of its 140 million population could be displaced.

Unfortunately if some regions are at risk from increased water levels, around one-third of the Earth's land surface is at risk of desertification, with over 250 million people directly affected and a billion people in over 100 countries at risk. According to the United Nations Convention to Combat Desertification (UNCCD), the world loses 24 billion tonnes of fertile soil every year. Failing arable land increases the likelihood that people will become displaced and increases the pressure on clean water supplies. By 2025, more than half the nations in the world are expected to face freshwater stress or shortages, and by 2050, as much as 75% of the world's population could face freshwater scarcity.

It is access to fuel, however, that sits at the centre of the debate about the interlinked nature of access to resources and potentially violent conflict. Energy security and the potential for an energy gap are becoming an issue for a growing number of countries. One example is China, the world's second largest oil consumer after the US (accounting for 40% of oil demand growth since 2000). While it has some oil reserves, it already imports about one-third of the oil it consumes. There have already been territorial disputes between Japan and China about oil and gas reserves in the East China Sea (which led to military confrontation), disputes with Vietnam about access to the Paracel Islands and oil exploration in the South China Sea. China's appetite for oil has led to increased investment in Africa, especially in the Niger Delta, where the lure of large oil reserves seem to outweigh concerns about human rights abuses in the region.

Oil and the economy

Oil is a finite resource on which the majority of economies depend. Fossil fuels, or hydrocarbon-based fuels, which are basically oil, natural gas (the vapour state of oil) and coal, have provided massive benefits in terms of cheap and powerful energy sources since the mid-18th century. Access to oil is critical to the modern economy – the majority of our products, machinery and systems are dependent on it and goods are transported around the world through the use of petrol, kerosene and diesel. The price of oil, given its fundamental importance, is often a critical factor in economic development. In the last couple of years, high oil prices have been fuelling inflation and causing immense problems for highly oil-dependent economies, particularly in the developing world.

The major emerging economies have been enjoying rapid economic growth, essential in enabling their populations to escape poverty. This has largely been powered by fossil fuels, however, increasing demand for oil and gas and exacerbating the scramble for scarce resources. The gradual shift of many governments towards a focus on carbon-free energy sources such as renewable or nuclear power has in many cases been driven by energy security, rather than climate change, concerns.

Indeed, the first real sign of a softening on the US front was in January 2007, when then President George W. Bush talked of his concern that America's reliance on oil was putting the country's energy security in the hands of foreign nations. This has been exacerbated by the enormous increase in the price of oil since 2005. While the price has fallen around $100 from its July 2008 peak of nearly $150 a barrel, long-term predictions are that oil prices will continue to rise over time. It is, of course, difficult to predict what will happen to the oil price in the next 10 years. Just as all market drivers looked set to drive the price over $200 a barrel in 2008, the global economy collapsed.

The prevailing sentiment had been that prices would continue to rise as strong global demand was driven by increasing demand from developing economies and slowing production due to geopolitical instability. Yet the economic crisis depressed demand, pushing down the price. The oil price has clearly fallen, however it does look as if the baseline has changed. Ten years ago anything over the price of $30 a barrel was excessive, but the baseline seems to have settled at twice that.

Of course, this is a short term issue. When demand returns, the oil price will soar again. In fact, with the economic crisis causing the oil majors to scale back investment in new capacity, it is possible that as demand recovers, supply will struggle to keep pace. There are things that we can to do to soften the impact of price volatility. By diversifying the energy supplies on which we rely, we can avoid an increasingly hostile scramble for resources in which we become ever more beholden to dominant energy suppliers. But when it comes to oil, our problems run far deeper than simply access to fuel.

Oil is a fundamental component of many of the goods that we take for granted. Some of the products that are made from oil range from crayons, fertilizer, plastics, make-up, ammonia, anaesthetics, antihistamines, aspirin, nappies, hearing aids, heart valves to perfume, vitamin capsules, shampoo, deodorants, dyes, batteries and more. We use oil for heating, cooling and transportation. Our current economic framework is completely dependent on oil, it's the engine of our society and, as developing economies grow, demand for those consumer goods we take for granted is also going to grow.

Peak oil

The concept of Peak oil – that global oil output will reach a peak and then decline – is the subject of ongoing debate. Given the increasing amount of energy required for, and difficulties in, extraction of oil, many believe it's not a question of whether it's going to happen but rather when. M. King Hubbert, the Shell geologist who came up with the concept of peak oil, said that the consumption of fossil fuel needed to be seen in historical context. He posited a bell curve showing a span of around 300 hundred years during which humans exploit fossil fuel – before its discovery they predominantly used wood, and afterwards – who knows?

Hubbert said, 'Our principal impediments at present are neither lack of energy or material resources nor of essential physical and biological knowledge. Our prinicpal constraints are cultural. During the last two centuries we have known nothing but exponential growth and in parallel we have developed what amounts to an exponential-growth culture so heavily dependent upon the continuance of exponential growth for its stability that is incapable of reckoning with problems of non-growth.'

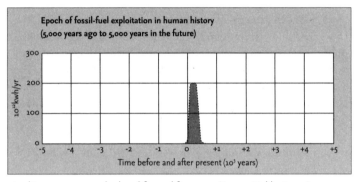

Looking at 5,000 years back and forward from 1974. (Source: Hubbert, 1974)

According to a global oil supply report presented by the Energy Watch Group at the Foreign Press Association in London in October 2007, world oil production peaked in 2006. Production will start to decline at a rate of several percent per year. Consequently, developing market demand will create imbalances that send oil prices up over time. In 2008, even the International Energy Agency (IEA) admitted that current trends in global supply and consumption are unsustainable. By 2020 and even more by 2030, global oil supply is expected to be dramatically lower. Some experts predict that oil should fall from 85 million barrels per day (mbp/d) to 18 mbp/d in 40 years and 8 mbp/d in 60 years.

Until the oil crisis of the 1970s, most people believed that energy was endless and that our use of it had no impact. In the early days of oil extraction, for every barrel of oil used in the exploration or extraction of oil, another 100 were discovered.

Even if there is oil to be found, it is becoming increasingly difficult to extract economically. One assumption is that our engineering capacity to extract oil is reaching its limits. That means it will take more money and energy to extract, refine and transport it until, at some point, when it takes the energy of a barrel of oil to extract a barrel of oil, further extraction becomes pointless, no matter what the price.

A US department of energy report commissioned in 2005 concluded that without timely mitigation, the economic, social and political costs of peak oil would be unprecedented. Our economic system is built on continuous growth, but that requires an easily available and relatively cheap supply of fossil fuels and it's becoming clear that we may no longer be able to rely on that. As detailed from the Peak Oil bell curve, it can be seen that:

1. the world will run out of accessible fossil fuels at some point fairly soon; and
2. these resources are likely to become more expensive until then

This means that we have to find ways of using alternatives to oil within our respective economies to kick start the beginning of a structural change within our economic systems.

Food security

While access to oil may be of more immediate economic interest, changes in weather patterns also have a direct and indirect impact on food supply. The retreat of Himalayan glaciers could mean a lack of drinking water and irrigation for millions. Without water, crops won't grow. At the same time, changing weather patterns, increased water salinity or desertification means lower food crop yields. Food supplies are increasingly being affected by the same access and pricing issues as oil and gas. Prices have been driven up by the increased cost of production at a farm level, in terms of fertilizer costs and equipment operation, through to transportation and shipping. Not only that, but we have even seen countries which have traditionally been major exporters of particular foodstuffs express concern about their own levels of production.

In early 2008, India, Vietnam, Argentina and China were among those countries that imposed export quotas to protect domestic commodity supplies amid rising concern over food production and price. Meanwhile, net food importers like the Philippines were left struggling to secure supply contracts when the production of food staples dipped below global consumption levels.

Given growing political instability, lower rates of food production, increasing costs of transportation and the potential diversion of foodstocks to fuel stocks (for biofuels – an issue I'll address below), the pertinent question seems to be how expensive is food going to get?

The first years of the 21st century saw spiralling food prices prompt government intervention in many countries. India, Morocco, China, Senegal and Indonesia cut import tariffs on wheat, while Ethiopia, Pakistan and Zambia imposed export quotas and, in extreme cases, export bans. Many countries, including Benin and Senegal, have introduced food subsidies in an attempt to keep basics at an affordable price, as did the Indonesian government following riots in 2006.

High food and fuel prices have sparked civil unrest in Central and West Africa and caused riots in Egypt, Mexico, Bengal and Yemen, amongst others. World cereal production has declined in the last couple of decades and not just in the developing world, mainly due to reduced plantings and adverse weather in some of the major producing and exporting countries, according to the International Food Policy Research Institute (IFPRI). Declines have been seen in both the US and the European Union and wheat production in Australia fell a staggering 52%, its cereal production by 33%. In January 2008, global wheat prices were 83% higher than the previous year and 2008 saw the export price for long-grain white rice reach $624 per tonne – in January 2001 it was $197 per tonne.

Of course, demand isn't purely for direct consumption or animal feed. In 2007 alone, world demand for coarse grain doubled due to the demand for grain used in US ethanol distilleries, according to calculations based on USDA figures by the Earth Policy Institute, an independent research organization. Separately, the IFPRI warns that if current biofuel targets are pursued, maize prices will increase by 72% and oilseeds by 44% by 2020. Shortfalls in production and subsequent price rises are being exacerbated by increasing competition for land, labour and water for biofuel production. In China, more than 3 million hectares have already been shifted out of rice production and into biofuel crop production.

It's not just cereals and grain that are coming under pressure. Increasing demand in Asian countries for meat and dairy products, due to higher incomes and changing diets – at a time when global grain production is at its lowest – is also pushing up prices. Heightened demand for meat products implies that greater tracts of arable land may be switched over to animal feed production, with further acreage shifted from crop production into livestock farming. On the other hand, the USDA predicts that higher grain prices, primarily due to the impact of crops being switched into ethanol production, will cause a decline in livestock farming as meat products become priced out of the market.

According to data gathered by the Goddard Institute for Space Studies (GISS), 2007 tied as the Earth's second warmest year in recorded history. Economist William Cline, senior fellow at the Institute for International Economics and the Centre for Global Development in Washington, DC, has estimated that global warming could cause a 16% decline in global agricultural gross domestic product (GDP) by 2020. He also projected agricultural output to fall by 20% in developing countries and by 6% in industrialized nations.

Rising temperatures have led to shorter picking seasons and more arid regions are already being tipped into the danger zone of drought, causing crop failure and loss of livestock. Australia, the world's second-largest wheat producer, is suffering its third consecutive year of drought. While slowing global demand for the crop has cut prices, they still remain significantly higher than in 2007.

Together, China and India produce more than half the world's wheat and rice, which makes projections by the IPCC regarding the possibility of the disappearance of Himalayan glaciers by 2035 all the more alarming. According to Lester Brown, founder of the Earth Policy Institute (EPI), Asia's melting glaciers may pose the biggest threat to food production the world has ever faced.

A series of reports, published by the Proceedings of the National Academy of Science (PNAS) in 2007 suggests that in coming years, rising temperatures and increased CO_2 levels could result in slightly higher crop yields in temperate regions. Meanwhile, farm-level adaptation measures may allow crops to cope with 1–2°C (33.8–35.6°F) of local temperature increases, basically 'buying time' to effect changes in carbon levels. Of course, any gains in productivity in temperate regions could well be cancelled out by agricultural decline in the tropics, with temperature increases between 1–2°C expected to cut rainfall and send staple crops over their survival thresholds. There would also probably be reduced livestock productivity and loss of cultivated areas in semi-arid and arid regions.

Even this scenario assumes the absence of extreme weather conditions which, over a few days could wipe out entire crops, if the conditions occur during a critical development stage. The reports note that this has already happened on a small scale, citing a case in the Po Valley in northern Italy, when extremely high temperatures in 2003 caused a record drop of 36% in corn yields.

Unfortunately the reports suggest that the variables used in simulation models to date have been oversimplified. In general, crop and pasture responses to climate change and corresponding increases in CO_2 remain largely unknown. Which means that predictions are likely to be fairly inaccurate. It is suggested though, that the overwhelming body of research indicates that any productivity gains won in the short-term regarding higher CO_2-related crop yields will be lost in the latter half of the century, as mean temperatures rise between 2° and 3°C (35.6 and 37.4°F) regionally and globally.

There are techniques that would increase the robustness of many production systems, including improving the conservation of soil and water resources, with techniques ranging from increasing crop rotation, use of genetically engineered crops, to the use of cover crops to conserve water and low impact tillage. The biggest problem, however, is that they tend to result in lower agricultural productivity – which is a problem in regions looking to increase food production.

What about Water?

Climate change is creating freshwater winners and losers among individuals, economies, societies and, of course, ecosystems and species. The reality is that a limited and somewhat fixed portion of the world's water is available and suitable for human consumption. Some 70% of the earth's surface consists of water, but only 3% of it is fresh water and less than a third of that is drinkable.

One of the great difficulties in understanding the impact of climate change on water systems is that weather, which obviously affects our water ecosystems, is for most of us a local phenomenon and we really only become conscious of it on a personal level – when we get caught in the rain because we forgot an umbrella. But melting glaciers and changes in the temperature of the oceans and land all affect precipitation, the source of almost all freshwater on earth.

'70% of the earth's surface consists of water, but only 3% of it is fresh water and less than a third of that is drinkable.'

The third edition of the UN's World Water Development report, *Water in a Changing World*, warns that water use is growing even faster than the global population. The 20th century may have seen a four-fold increase in population but water use has tripled in the last 50 years alone, and the population is growing by around 88 million a year. The amount of water we consume is increasing, whereas the supply of freshwater is static. More than a third of the world is already under water stress, a situation where poor water supplies affect food production, economic development and even human health. The situation is worsening and if we add 3 billion people to the earth by 2050, there won't be any increase in the amount of water available to us.

The report warns that water shortages are already beginning to affect economic growth in California, China, Australia, India and Indonesia and warns that conflicts over access to water could break out in a number of countries, including Haiti, the Middle East, Sri Lanka and more. We've discussed the fact that rising temperatures could result in glacier melt in the Himalayas, which have the largest concentration of glaciers outside the polar caps. According to the WWF, 67% of the glaciers are already retreating, causing glacial lakes to burst their banks, destroying villages, crops and livestock. At the UN climate change conference in Bali in December 2007, Bhutan pleaded for adaptation funds to help protect its glacial lakes from bursting. Annual glacial melt is necessary for crop irrigation during the dry season in both China and India. And the glaciers feed the Yangtze and Yellow rivers and provide up to 70% of water in the Ganges in India during dry season. If the glaciers melt there will be no slow release of water for irrigation in any of these regions.

Increasing population, industry, economic growth and urbanization all put additional stress on the provision of clean water from dwindling reserves. As well as stretching renewable supplies of fresh water to their limits many countries are severely depleting the finite supply of water from underground aquifers. The UN estimates that 1.2 billion people worldwide do not have access to safe drinking water and the OECD estimates that around 5.2 million people die each year from waterborne diseases, caused, in part, by lack of proper sanitation and treatment of wastewater.

WWF has called for a global water agreement, specifically to help govern and mitigate conflicts with waterways that conveniently form international borders. If brought into force and widely implemented by the nations sharing the water of river systems and associated lakes and aquifers, the convention could greatly contribute to ending the chaos of water grabbing and to improving the health of 263 rivers and lakes in 145 countries. Rivers that cross or form borders, most suffering from nonexistent or inadequate regulation, drain half the earth's surface, provide water to 40% of the human population and generate about 60% of global freshwater flow.

Increasing areas of the developing world are coming under severe water stress, whether through land-use change, pollution or drought. In the developed world, however, concern has predominantly been focused on the need for sufficient water resources to serve our industrial base. Now these concerns are expanding, as many regions move beyond such problems into severe water management issues, especially in the US and Australia.

California has been suffering from low rainfall for a number of years, which has led to water supply problems. Around two-thirds of the State's precipitation occurs in Northern California, while two-thirds of the population live in Southern California – with most surface water in rivers and lakes derived from melting snow in the Sierra Nevada mountains. There have been changes in mean winter temperatures, which has affected the accumulation of snowfall in the Sierra Nevada mountains, directly impacting on the water available. At the same time, much of the water consumed in the South must be transported from the North. This water transport is highly energy intensive – water transportation, storage and treatment are understood to account for nearly 20% of the state's electricity consumption.

The combination of a rapidly growing economy and population with diminishing water supplies means that the issue is becoming critical. During the 2008–09 winter the State warned that water deliveries could fall by up to 30%. While individuals may not see the difference in the winter climate in the mountains, they're very much affected by its results. This can range from something as simple as needing to change garden planting from thirsty grasses to plants that can survive long periods without watering, to something as drastic as increasing wildfires. And it's not just individuals – California has a world famous wine industry and there are growing concerns that, driven by the need for water, many vineyards may have to move further north into Oregon or Washington. All of these impacts are a result of trends in California's climate that are likely to continue in coming decades.

In the developed world we need to adapt to new water requirements, ranging from the reduction of water consumption to compensate for reduced precipitation rates, to shifting the location of an industry away from a drought-prone area to one expected to receive higher water flows. Perhaps the greatest short-term threat from climate change to freshwater ecosystems is the interaction between problems such as over abstraction and misallocation (such as when a golf club is kept green rather than farmland irrigated), and the growing habitat fragmentation of vegetation into small sections, which can rapidly affect the ability of the land to manage and maintain freshwater systems.

Of course, there's the other obvious problem – that of conspicuous consumption. Some 70% of the quality drinking water flowing into North American or European homes is flushed away or used for cleaning. People consume water through drinking, cooking and washing but far more is used in the production of food, cotton clothes, consumer goods and other trappings of modern living. Part of the problem is that the consumption of water at the point of production is not priced into consumption of the resultant goods, and this can lead to enormous inequities in the water balance.

A critical measure that the WWF *Living Planet* report added in 2008 was global, national and individual water footprints – a measurement of the direct and indirect water consumption by goods and services in different nations. This shows up the significance of water traded in the form of commodities – such as a simple cotton T-shirt, which requires 2,900 litres of water in its production. On average, each person consumes 1.24 million litres (about half an Olympic swimming pool) of water a year, but this can vary dramatically from 2.48 million litres annually per person a year (US) to 619,000 litres (Yemen). This is an unsustainable consumption pattern and there is insufficient water to support it.

Many of the products consumed in the developed world are produced in the developing world, which means that the developed world is consuming developing world water resources without having to pay for them. According to the 2008 *Living Planet* report, Japan has a water footprint of 1150 cubic meters per year per capita, but has about 65% of its total water footprint outside the borders of the country. By comparison the water footprint of China is about 700 cubic metres per year per capita but only 7% of the Chinese water footprint falls outside China.

According to a report from Ceres and the Pacific Institute, 10 out of the world's 14 largest semiconductor companies are based in the Asia-Pacific. Water is a vital component of the manufacturing process for semiconductors – in 2007 Intel and Texas Instruments used 11 billion gallons of water in manufacturing silicon chips. If a lack of usable water leads directly to a collapse in semiconductor manufacturing, how long is it likely to be before this has a direct impact on the computer and electronic goods market in the developed world?

If the effects of climate change continue, degradation of freshwater systems will accelerate ecosystem degradation. This could result in further reductions in the production of food, goods and services, and this will have a direct economic and geopolitical impact.

The 'real' cost of carbon

In order to evaluate the real cost of carbon, we need to understand the impact of climate change on aspects of our environment to which we have never set a value.

If we purchase a piece of electronic equipment from China, with the cost of alleviating pollution, climate change, energy and water consumption factored in the price, that price would probably be far higher than what we pay today. Today's economic framework seems to focus on an immediate cash price, with no pricing mechanism for social, environmental or long-term capital impacts.

One area where attempts have been made to understand the actual cost of the decline of natural resources has been with regard to the value of forests, on behalf of the EU. In 2007 the EU commissioned a report, The Economics of Ecosystems and Biodiversity (TEEB), on the impact of forest loss on the global economy. The first phase concluded in May 2008, when the team released interim findings that forest decline could be costing about 7% of global GDP. The second phase expands the scope to other natural systems and is due for delivery in 2010.

Changes in land-use, climate, pollution and water use can lead to changes in biodiversity and ecosystem functions. This, in turn, leads to changes in available ecosystem services and, therefore, a change in the economic value. It is the assigning of specific values to these benefits that enables us to set a price for their loss.

The critical message of the TEEB report is that as forest cover declines, there is a fall in the ecosystem services it provides, as clean water, fertile soil and carbon storage opportunities decline. While alternatives can be created by human activity, through the development of water reservoirs, farming using genetically modified organisms or through building carbon capture and storage facilities, this can carry a significant cost and we may not be able to predict the consequences. Headed by Deutsche Bank economist Pavan Sukhdev, the report puts the annual economic cost of losing forest at between US$2 and 5 trillion per year.

For much of the developed world, the concern that we might be losing an important carbon storage opportunity seems a distant problem and one that can't compete with the critical immediacy of banking problems. But if we fail to protect that opportunity, the costs of finding an alternative are only likely to increase. Either way, there's a financial cost.

Outlining the long-term consequences of economic, political and social action helps to ascribe value to aspects of our environment which have been ignored. In a capitalist society and, despite its recent hiccups, capitalism is the dominant political and economic theory; things without a value ascribed to them are usually ignored.

A number of nations, businesses and global organizations are already exploring different ways to finance forest conservation and there are signs of a trade in natural ecosystems developing. In fact, one of the great victories of the UN climate change conference in Bali in December 2007, was an agreement to explore how forestry could be added to the market framework under the Kyoto Protocol (which we will talk about in more detail in Chapter 5). There is an increasing acceptance that financial market models can be used to calculate value, and introduce that price into the wider economy.

'In a capitalist society things without a value ascribed to them are usually ignored.'

If we are able to restructure our approach to what we value, then we may have a way to overcome the political problems inherent in the difference between long-term and short-term benefit. Assigning a price to carbon encourages behaviour away from the use of fossil-based commodities, either in the generating of energy, the powering of transportation or the manufacturing of goods. If we are successful in that, it could be the first step towards quantifying the externalities in current goods and services, and pricing our decisions accordingly.

Driving Change

3

The obvious response to recognizing the cost of our behaviour is to change it. Human nature being what it is, however, that's not always the response that we get. What we need to do is find an efficient and cost-effective framework for encouraging that change in behaviour.

Risk versus opportunity

Cutting carbon emissions is about risk management. There is no absolute certainty that temperatures will rise to dangerous levels but there is a strong probability. Action on climate change is about cutting the chances of a dramatic degradation of planetary resources from 50–90% to less than 5%. Given the growing population and an awareness that existing resources of energy, clean water and fertile land will, at best, remain stable, managing that risk becomes critical.

We know that the long term costs of climate change are going to be high but while fear for the future can be a factor in motivating structural change, the reality is that most of us have more immediate concerns. What may well define our response to climate change is our ability to balance the risk and opportunities available.

While there will be a significant cost in achieving change, there are also potential new markets and opportunities that will arise, and these extend far beyond the energy industry alone. While power companies and energy-intensive industries may have been the initial targets of carbon emissions cuts, it is clear that to successfully achieve necessary emission reductions, limits must be implemented throughout industry and the wider economy.

Cost effective solutions

In a modern society, it is the economic path chosen by industry that has the strongest impact on the technologies and products that we consume. One of the most essential aspects of how we attempt to achieve a transition to a low carbon economy is going to be the cost-effectiveness of the processes or technologies that we use. In 2007, Swedish power company Vattenfall asked consultants McKinsey to work with them at exploring the likely cost of the abatement of carbon: the resulting report, A Cost Curve for Greenhouse Gas Reduction, explored different GHG reduction strategies. Simply put, they worked out what the different approaches to cutting carbon would cost and displayed it as the 'Vattenfall Abatement Curve' (see page 57).

The measures have been arranged in order of cost, with the cheapest on the left and the most expensive on the right. Only measures with an estimated cost of less than €40/tCO2e were included at the time of the original analysis. What was most interesting was that they discovered a quarter of all possible emissions reductions would have no net cost. While there might be initial expenditure, as with the use of energy efficiency technologies, the action would save more money than it cost over the medium- to long-term.

Many of these are activities that increase efficiency, especially with regard to buildings (including insulation, heating, air-conditioning) and vehicles. According to the report, about 7 gigatonnes (Gt) of CO2e emissions per year could be avoided at a negative cost of abatement. Beyond that, nuclear power, livestock management and reforestation are the lowest cost options.

Avoided deforestation (stopping the destruction of existing forest, such as the Amazon) looks fairly expensive, but has far-reaching consequences that should change the way we ascribe value to it, in terms of ecosystem support, biodiversity and even water management. And its worth remembering that the prevention of deforestation and

Vattenfall Abatement Curve

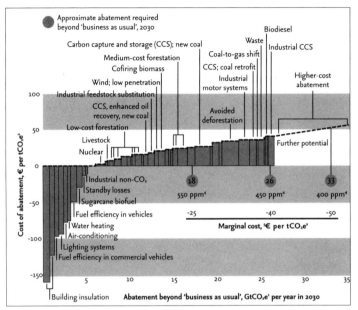

(Source: Enkvist et al, McKinsey & Company, 2007)

The abatement cost curve describes two numbers:

1. The overall potential to reduce CO2 equivalent (CO2e) emissions.

2. How much that measure costs for every tonne of CO2e emissions it saves.

peatland destruction requires no technological development and little capital investment for the activities themselves – important factors as we enter a prolonged economic downturn. The economics are surprising – it has been argued that if developed countries spent the same amount of money on preventing deforestation and the destruction of peatlands as they do on biofuel subsidies ($15 billion in 2007), this could halve the total costs of tackling climate change.

The McKinsey analysts pointed out that the projected cost of cutting CO2e levels to 450ppm could be only 1.4% of global GDP in 2030, or €500 billion. Cutting emissions to 400ppm would only cost

1.8% of global GDP in 2030. Obviously, if the target is more stringent or the more expensive approaches selected, then the cost will be higher. But at this level the cost is relatively small – in 2005 global spend on insurance alone was 3.3% of global GDP in 2005. Given that the amount spent on insurance is likely to increase as companies have to insure themselves against the impact of climate change, might it not prove cheaper just to try to solve the problem?

While no single technology or solution can solve the problem on its own, they certainly have impressive potential when combined. The cost curve shows how we could reduce annual emissions by almost 27 billion tonnes spread over the entire world and every sector of the economy. And, as low cost opportunities are crucial for the world to be able to meet the necessary cuts at a manageable cost, no region or sector can be neglected.

It's important to remember that these costs were set out in terms of a goal of cutting emissions by 50% by 2050. They were based on the belief at the time that the decision was between keeping emissions levels at a range between 550–450ppm. In early 2009, measurements taken by Stockholm University reported that average concentration of CO_2 rose from 390 to 392ppm in 2008. Combine this with other GHGs and it's possible that we're already close to emissions levels of 450ppm.

The IPCC has said that emissions should peak by 2015 to avoid the worst impact of climate change, limiting warming to 2–2.4°C (35.6–36.3°F). Many have interpreted this to mean that emissions need to be cut by around 40% by 2020 and between 60–80% by 2050. This means the question of how and how fast we should cut emissions remains open to debate and it's clear that the further and faster we need to cut, the higher the cost is likely to be.

That cost is going to have an economy-wide impact. For the majority of companies operating in traditional industries, environmental issues are now presenting them with material risks, ranging from increasing costs of energy, commodities and manufacturing, transportation and distribution, all the way to the impact of their behaviour on brand and reputation. At the same time, the potential for new industrial development, industries and market positioning is leading to dramatic changes in business planning.

The launch of CO_2e emissions regulation under the EU's Emissions Trading Scheme 2005 (see Chapter 5) created an economic risk factor for business which previously hadn't existed. It was a pivotal moment in terms of an understanding of the need to measure, monitor and manage emissions. Under the new standards, those installations that failed to meet emissions limits would be forced to buy emissions credits or risk being fined.

Increasingly, investors and shareholders are now taking account of the need to factor climate change issues into company valuations, as the potential impact will vary from industry to industry. Both the UK-headquartered Carbon Disclosure Project (CDP – representing investors with over \$57 trillion under management) and the US-headquartered Investor Network on Climate Risk (INCR – investors with over \$7 trillion under management) are groups focused on the importance of understanding the financial risks and investment opportunities posed by climate change. They pressure corporations to report consistently on carbon emissions within their business, as the first step towards managing them. Shareholders are putting increasing pressure on corporations to focus on the green agenda, not because they are concerned about climate change of itself, but because ignoring the potential risks would be fiscally irresponsible. Investors want companies to be aware of the business impacts of climate change.

A tipping point for the awareness of the role of investors in shaping company policy was the Exxon Mobil shareholder revolt, which occurred in early 2008. Descendants of the company's founder, famous oil man John D. Rockefeller, led an attempt to restructure Exxon's management and force a focus on the renewable energy market. Historically, Exxon had been an outspoken opponent of any US national climate policies to cut GHG emissions, while competitors Chevron, BP and Shell had invested billions of dollars in renewable energy and set company-wide GHG reduction targets.

The shareholder revolt saw 19 institutional investors with 91 million shares worth $8.6 billion table a motion asking Exxon to address climate change risks and opportunities. The group weren't prompted by concerns that Exxon wasn't making enough money: in 2007 the company made around $40bn in profit. Instead, it argued that Exxon, by focusing solely on the oil markets, was failing to address the inevitable changes likely to be driven by the changing role of oil in a low carbon economy. While a majority of shareholders subsequently blocked the motion, the fact that the move was made at all shows how the early 21st century has seen a major shift in the way the energy and the economic environment are perceived by the investor community. This was a direct attack on the business model of a global oil giant.

A question of economics

The early 21st-century global financial crisis has naturally led some politicians to question the affordability of proposed climate change goals and whether the transition to a low carbon economy is anything more than an optional extra in a time of financial crisis. Others believe a green agenda provides an opportunity to fix financial and climate problems at the same time. The reality is that a focus on solving climate problems may be the only real option we have to move forward. Our current economic framework doesn't place any significant value on the natural resources of our world – it undervalues future wealth or poverty in favour of 'now', which has led to a focus on the short-term extraction and consumption of existing resources, severe environmental degradation and an outpouring of GHGs into our atmosphere.

Much of the commentary about the global economic crisis has talked about how the short-term drive for profit, led investors around the world to take poorly-understood risks as to where they placed their money. It's this short-term view that underlies many of the problems that the low carbon agenda is trying to reverse. This is not only a question of whether we act in a morally correct fashion, or take a particular political or philosophical stance. It's actually a question of whether economies can afford not to take the green, low carbon route.

Economists such as Nicholas Stern have warned that the consequences of ignoring climate change would be far greater than the consequences of ignoring potential risks to the financial system. As different countries shift from deploying economic stabilization measures to promoting spending to drive the 'real' economy, there is an opportunity to bring about a new economic framework. Stern suggests that increased public spending on public transport, energy and green technologies would boost the economy.

> 'Our current economic framework doesn't place any significant value on the natural resources of our world – it undervalues future wealth or poverty in favour of "now"'

The carbon and environmental markets could provide a route out of financial crisis. It certainly seems obvious that in times of geopolitical instability developing local power supplies is a sensible choice. In October 2008, British Prime Minister Gordon Brown said: 'The climate change agenda is part of the solution for many of the problems we face as a global economy'. He stated that Europe's move to a low carbon economy could help reduce costly and volatile oil imports and protect it from sudden disruptions to supply.

There is no question that faced with a gloomy economic outlook, it could be easier for legislators and policymakers to prioritize short-term economic considerations and keep the extra costs associated with climate-change legislation away from their domestic industries. But easier doesn't mean that it's the right choice to make.

A Green New Deal

After the 1929 Wall Street Crash which adversely affected the US and economies around the world, US President Franklin D. Roosevelt launched the famous New Deal federal programme, in the 1930s, which focused on creating jobs, reforming business practices and promoting economic recovery. Many experts believe that the solution to our current economic woes could be solved by a 'Green New Deal', a programme of public expenditure on low carbon infrastructure projects, combined with smart market regulation to kick-start the economy, create jobs and promote economic growth.

In 2008, the UNEP launched the Green Economic Initiative, a three-pillar plan focused on the need to direct nature's resources into national economies, to generate employment and to lay out robust green policies and market mechanisms to accelerate the transition to a low carbon economy. It called on world leaders to put money into job-creation and new technologies, diverting it away from the speculation that lay underneath the traditional global banking system. UNEP executive director Achim Steiner said that a worldwide banking bailout had been mobilized in only a few weeks, while the response to climate change remained slow. He warned that from 1981 to 2005 the global economy more than doubled, but that 60% of the world's ecosystems, including fisheries and forests, were either degraded or suffered over use.

The UNEP research also identified the emergence of a green economy, which it expects to create millions of jobs in the next decades. Whether creating clean-burning fuels, installing solar water heaters, or improving the energy efficiency of homes and offices, new jobs will form a key part of the market for environmental products and services, which is expected to reach $2.7 trillion by 2020. In the UK, the New Economics Foundation similarly launched the concept of a Green New Deal. In the US, President Barack Obama's promise of investment into the renewable sector has also been largely driven by the new industries developing, which can create economic growth and employment.

We have have an opportunity to reassess how we wish to live our lives. Our choice is simple – we can create economic opportunities in new industries, develop newly energy efficient products, develop a secure network of distributed energy generation, smart and efficient energy grids and transform our economy and our way of life. Or we can choose not to, and reap the consequences.

Given that emissions into the atmosphere have a global impact, at the very least, we should want an international agreement that ensures we're all working towards the same goal. In order to reach this agreement however, we may well need to explore a number of approaches to encouraging lower carbon activities across entire economies.

Changing Behaviour

The most effective methods to manage carbon emissions are fairly well understood. Forum for the Future calls it the ARRO hierarchy, for *Avoid, Reduce, Replace, Offset* (see diagram below). There are an array of tools that can be used to encourage people to undertake this form of behaviour. These can range from incentives to encourage the adoption of renewable energy such as a feed-in-tariff, tax breaks for the deployment of specific technologies (such as the Production Tax Credit in the US or capital allowances in the UK), subsidies, agreed standards for energy efficiency or building developments, investment in research and development, even technology transfer.

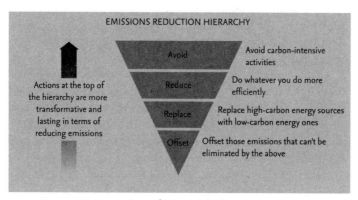

(Source: Forum for Future's *Getting to Zero:*
Defining Corporate Carbon Neutrality, June 2008)

While there are definite benefits to utilising all these different approaches, in the end, in order to effect an economy wide change in behaviour there are really only two options that will significantly affect the emissions of GHGs:

- command and control
- setting a price for carbon

Command and control

Command and control is a policy approach where polluters, in this case emitters of CO_2e, are given highly specific regulations ensuring control of the selected pollutants, and are often prescribed alternative technologies to use. Effectively, it's about punishing the emitter for their behaviour.

The most obvious way to limit a specific pollutant is to ban it and governments can take such action if they choose. Traditionally, pollution has been controlled with a range of different approaches, from the outright banning of certain chemicals, to limits on allowed discharges. This is usually accompanied by the monitoring and measurement of amounts of a pollutant, regulated by independent bodies and with stiff fines for contravention.

For example, the 1956 Clean Air Act in the UK was passed in response to the Great London Smog of 1952, in which a combination of adverse weather conditions and smoking fires from homes and industrial facilities led to about 12,000 deaths in excess of natural mortality rates, from acute and ongoing effects of the smog. Following the smog, a report from the Beaver Committee suggested that smokeless zones be introduced into London, meaning that no coal or wood could be burnt, either domestically or commercially. This formed the crux of the Clean Air Act and directly led to the clean-up of London air.

There can be problems in implementing an outright ban, however, especially when industry believes the banned pollutant is critical to the operation of their business – it is never an easy move for a politician to make. For example, even the introduction of limits to CO_2 emissions under the EU ETS caused cries of outrage from industry, amid fears that the cost would be so high as to make their operations internationally un-competitive. Intensive lobbying by industry resulted in an agreement that saw initial permits to pollute issued to installations for free. It was suggested that this was in order to kick start the market – rather negating the idea that the introduction of emissions permits would cause emissions to become more expensive. It is clear though, that any attempt to ban the emission of CO_2e overnight would be doomed to failure, as the entire economic infrastructure ground to a halt. Virtually all human activities emit some form of CO_2e.

One of the problems with the command and control approach is that it is likely to be less cost-effective than other options. This is partly due to the fact that the prescriptive selection by government or regulators of specific technologies may not be the most effective, as it doesn't allow prices to come down through market forces and doesn't allow for innovation within that market. The difficulty lies in working out the marginal abatement cost curve (MAC curve) – the marginal cost of reducing pollution or emissions. This differs from industry to industry, depending on their product and technology use and is especially complicated on the issue of GHGs, as there are a range of gases being emitted in almost every area of society and the economy.

Setting a price for carbon

There are two core approaches to setting a price for pollutants: taxation or a cap-and-trade programme. These are considered to be more cost-effective and efficient than command and control in cutting emissions, as they generate a price for the pollutant to be avoided – in this case, a carbon price. This enables the private sector to plan for the cost of emissions and, as long as the cost of emission remains higher than the cost of pollution, encourages investment in alternatives, stimulating private sector engagement in changing pollution levels.

A tax sets a direct fee (the carbon price) on either the carbon emitted through the generation of power or the carbon embedded in the development, manufacture and distribution of a product. What it doesn't do is set a limit on how much of the pollutant is emitted. The cap-and-trade approach sets a limit (a cap) on the amount of pollutant that can be released, either by an industrial sector, a country or a region.

Allowances, or credits, are issued equivalent to that fixed amount of pollutant and these can either be used by polluters or sold on by those who've managed to cut down their own emissions. What a cap-and-trade scheme won't do in relation to a carbon trading scheme is set a specific price for carbon but rather it will let the price be determined by the supply and demand of allowances in an emissions trading market. There are positive and negative aspects to both approaches.

Carbon tax

A tax is perhaps the clearest way to introduce a price for carbon, as the price is immediately obvious. Such a tax could be levied upstream on the producers of fossil fuels, at the point of emission on industrials, or at the point of consumption on consumers.

At first glance, a carbon tax at the consumer level seems like the most effective way to add the environmental cost of emission to the cost of the product or service. The external harm (or externality) created by the carbon emissions is added to the cost of the product, resulting in the true cost. A clear benefit of an upstream tax, however, is that the carbon content of fossil fuels, from anthracite to lignite coal, oil and natural gas, are known. This means that the amount of CO_2 released into the atmosphere through the extraction and combustion of such fuels would be easy to calculate, presenting few problems with documentation or measurement.

Such a tax would obviously be felt throughout the supply chain. By taxing fuel and power sources according to their carbon content, the cost of that choice could be clear at every stage of the value chain: from making decisions about travel arrangements, the build or purchase of appliances or housing, business decisions on product design to decisions about investments and/or energy and facilities management, consumers would be clear about the carbon impact of their decisions.

If a carbon tax was imposed on businesses in general, rather than the energy industry, the impact on each company or industry sector would depend on how much fossil-fuel-based energy it uses, how higher energy prices affect their business and the company's ability to either minimize or avoid increasing costs (e.g. by using fuel more efficiently or using cleaner fuels) or pass along costs to its customers.

The reality of implementing such a tax would be easier in theory than practice, as we have problems with ascertaining the real external cost of CO2e, as well as with clearly allocating responsibility for emissions. At the same time, tax is a sovereign national issue and any attempt to create, implement and collect tax on a global scale would be enormously complex.

One of the arguments against the introduction of a carbon tax at a national level is the problem of competitiveness. Proponents of such taxes argue that either companies would fail to compete against their rivals or that the tax would encourage innovation and new efficiencies. Many industries have concerns regarding carbon leakage, which is where an industry moves from a region with tight carbon regulation to one without such a regime. Not only would this undermine local and national industry but there would be no net environmental benefit.

At the same time, analysts have found it difficult to define exactly what impact such a tax would have on emissions. Achieving the necessary reductions would depend on whether the carbon tax raised prices to a point that significantly cut consumer demand, while promoting the development of low carbon technologies. The difficulty lies in gauging market response because that can frequently vary for different goods and services. The increase in US gasoline prices in 2008 was a prime example of how cost can influence consumer demand. The price of $4 per gallon (between $2–$3 in 2005) resulted in Americans driving less. Yet it's difficult to estimate what the response would be to a similar rise in electricity prices, when that is such an integral part of everyday life.

There are other issues, such as how to deal with international trade, if the tax is not equal on all nations. Imposing taxes on imports could lead to trade wars, as countries result to tit-for-tat tactics: In 2008, as the EU was debating its own climate change legislation, there was discussion about the imposition of a carbon tax on imported goods from outside the EU. This caused outcries from non-EU countries such as the US and China. Even if a tax was imposed internationally, setting fair levels of taxation for countries with enormously different economic power is likely to prove difficult.

The concept of a carbon tax has also been criticized because of its potentially detrimental impact on the poor or disenfranchized who, as a general rule, spend a greater percentage of their income on fuel. Even if the tax was on goods, not fuel, those in a lower income bracket would immediately be paying a higher proportion of their income for those products and services. The introduction of a 'revenue-neutral tax' might solve this (a tax from which the government doesn't make any money, and the funds raised are returned to the people through rebates, possibly even on a per capita basis). Another option is a 'tax-shift' where, in order to smooth the transition of tax, during an initial period of the implementation of a carbon tax, other taxes would be phased out.

The biggest problem with a carbon tax is that it doesn't set a limit on emissions. It simply adds a selected cost of carbon to the products that we use. If we added the cost of mitigating carbon to the operational costs of all energy producers and oil refiners at source, we may well cut down the amount of carbon in the economy, but at a realistic cost we could price such energy production out of the market. While this may be beneficial over the longer term, its unlikely that the existing market could function under such a dramatic change. If we are to alter the status quo, we need to find a way to provide both a cost and an opportunity. While we need to effect a transition to a low carbon economy, we need to do it in such a way as to support as cost-effective a transition as possible.

Cap-and-trade

One way to do this is to use a market mechanism such as cap-and-trade, which was set up in the Kyoto Protocol (see Chapter 5). One of the key differences from taking a taxation approach is that emissions would be limited to an agreed cap. Each entity regulated under such a scheme, be it a country, company, installation or industry sector, would be responsible for measuring, monitoring and reporting its emissions. At the end of each compliance period (the period of time over which each entity must comply with the targets), the emissions of each party must meet or be below its allowance. If not, it must buy credits to cover the excess emissions, from another party who has managed their emissions allowances more effectively.

The intention of any cap-and-trade scheme is to limit the total amount of allowances available. This scarcity creates a market price for the allowances based on supply and demand. If the emission of CO_2 adds cost to doing business, then the price of credits allowing the emission of CO_2 should settle at that cost. Regulated emitters may buy and sell allowances; this means those companies which can cheaply or easily reduce emissions can sell allowances to other companies for which such reductions are more expensive or difficult. This flexibility enables businesses to make changes to their operations and manage the financial impact of the cost of carbon in the most effective and cost-efficient way.

The cap-and-trade approach does have its problems. Under such a scheme it's possible that if the price of credits remains low, it will be more cost-effective for industry to emit GHGs than avoid it, negating the point of such a scheme in the first place. One of the reasons that a cap-and-trade scheme is much beloved of free marketers and politicians though, is that it responds to the market and provides a degree of flexibility. While it could prove difficult to maintain a carbon tax during a recession, theoretically a cap-and-trade programme will respond to falling demand (less use of power as manufacturing output falls) and the price of emissions allowances will fall. This was certainly the case within the EU's ETS in 2008–09.

The price of carbon under such a scheme responds to a combination of emissions allowances released and market demand. That means that the initial level of the emissions cap and how it is adjusted over time are critical in influencing the price of allowances. Similarly, the allocation of allowances, point of regulation, cost containment mechanisms and provisions for offsets can all influence how successful a cap-and-trade programme can be.

The point about the carbon price within a market mechanism is that it responds to market signals. If the price of a credit falls below the level where alternative behaviour is rewarded that means one of two things has happened. Either emissions have fallen so dramatically that the credits are not required, or that the cap on emissions has been miscalculated – either way the requirement for credits is insufficient to maintain a sufficient price. What this tells us is not that the mechanism itself has failed, but rather that the framework within which it operates is ineffective.

It's worth discussing here the most controversial aspect of a cap-and-trade scheme, and that's the use of offsets. This is where regulated installations can meet their carbon allowance requirement through the use of credits generated outside its own actions. Emissions that cannot easily be cut can be balanced by either the removal of GHG emissions elsewhere, or the replacement of a high emissions project with a project with lower or no emissions under another, connected, regime. For many, offset has become a term with negative connotations. Environmentalists, for example, fear that it is no more than a means for the developed world to brush off their responsibility to cut emissions onto the developing world.

The important thing is that it is emissions into the atmosphere on a global basis that count. It's worth remembering that the term offset means to 'make up for' or 'compensation'. The climate challenge is a challenge of scarcity – we need to reduce absolute emissions. Offsets may look like they are offering a free lunch but that's not what they're about – the price of the offset comes from scarcity and that can only come from an obligation to reduce. It's this price of the offset that is critical to changing people's behaviour.

That means that as long as a market for offsets exists under an obligation to reduce emissions, and when offsetting (done in a structured and regulated fashion) is predominantly used as a means of easing the transition to a low carbon economy, it has a valid role to play in that transformation.

There is no question that, as and when economic activity picks up, demand for credits under the current schemes are going to rise significantly, which will have a direct impact on the price. Crucially, if we want to use a market mechanism to set a carbon price, we need a detailed understanding of the true cost of emissions over the short-medium- and long term. Only then can we set the appropriate emissions targets, as well as caps across all industry and end the gifting of allowances, which will enable the cap-and-trade scheme to reflect the market environment accurately.

Understanding emissions

The question of how we allocate responsibility for emissions is becoming increasingly important. Alarm regarding national and industrial competitiveness and carbon leakage are likely to cause significant problems in reaching a global agreement post 2012. This means that understanding where GHGs are emitted in the supply chain is a critical component in our response to climate change. For example, according to research from Carnegie Mellon University, 33% of China's territorial emissions come from the production of exports. Half of China's emissions growth between 2002 and 2005 were due to export production; another third of that growth came from capital investments, with a significant share of this in export industries. Indeed, only 15% of emissions growth between 2002 and 2005 was due to household and government consumption.

Acid Rain Programme and the US Sulphur Dioxide Emission Trading

The cap and trade of sulphur dioxide (SO_2) in the US coal industry is credited with not only cutting SO_2 emissions by 50%, but doing it at a fraction of the projected cost.

Since the 1960s SO_2, which reacts in the atmosphere to produce acid rain, has been viewed as hazardous to the environment and human health. The US Congress opposed a number of bills intended to cut emissions in the 1980s but finally the concept of sulphur trading was agreed upon and eventually introduced in 1994.

The scheme proposed a reduction of annual SO_2 emissions by 10 million tonnes from the 1980 level – about 50% between 1980 and 2010, according to the Environment Protection Agency (EPA). The EPA Clean Air Markets Program also provided a market-based regulatory programme. This was one of several different market-based mechanisms that used a variety of economic incentives and disincentives, such as tax credits, emissions fees, and emissions trading. The energy industry was horrified and lobbyists warned it would cost $10 billion a year and decimate the industrial base through the financial impact. In fact, the actual overall cost of the entire programme came in at around $1 billion.

While 1995 was the first year that affected utilities were required to achieve compliance with SO_2 emission limits, the regulations that established the programme required EPA to offer annual auctions of allowances starting in 1993. The first auctions, held in March 1993, made available 50,000 of the 1995 allowances and 100,000 of the 2000 allowances. The clearing price for the 1995 allowances was $131. Despite claims by lobbyists that the cost could go as high as $400, the average cost of a credit has remained around $150 in the early years of the scheme.

The flexibility that trading gave to utilities helped reduce costs, but other factors were also important. Scrubbers to remove sulphur from power plant smokestacks turned out to be far cheaper to install and to run than had been anticipated and rail-freight deregulation sharply reduced the cost of transporting coal from one area of the country to another as rail prices fell.

The lesson from the US sulphur market is simple. By putting a price on carbon, we can limit its use, drive investment into alternative technologies and cut the potential impact of emissions, probably far more cheaply than we expect.

In March 2009, US Energy Secretary Steven Chu warned that the US may look at imposing carbon import taxes as a means of ensuring the US remains competitive under a federal cap-and-trade scheme. Responding to Chu's comments Gao Li, director of China's Department of Climate Change, said that countries that buy Chinese goods should be responsible for CO_2 emitted by the factories that make them in any global plan to reduce GHGs.

The national levels with which we are presented often give an unrealistic and potentially unfair picture of emissions growth. We need to find better ways of tracking emissions in order to address them.

Measure to manage

Obviously the need to measure emissions is necessary beyond simply underpinning decisions about policy frameworks. If we're going to be able to avoid, reduce, replace and offset emissions, we need to be able to work out where to cut, what technologies to use and how we should transform our manufacturing and distribution processes.

The first step has got to be finding out where our current emissions come from and how best to measure them. Then we need to agree where we are going to measure emissions – either at the point of fuel extraction, power generation, manufacturing, transportation or consumption? It is common practice in most industrial environments to monitor, measure and control the discharge of various pollutants, but before the Kyoto Protocol was agreed, CO_2 emissions were not considered.

Initial attempts to measure CO_2e emissions focused on fuel consumption, as it's relatively simple for scientists to work out how much CO_2 is emitted per tonne of fuel incinerated. When the fuel is combusted, there's a reaction and the carbon in the fuel reacts with the oxygen and creates CO_2. This is basic chemistry and the rate of gas generation is fairly well understood. It's also fairly straightfoward to manage those emissions, by cutting down fuel consumption.

When one moves beyond power-related emissions however, new methodologies have had to be developed. Direct emissions from power and fuel consumption are relatively easy to measure, but the process becomes far more complex as we attempt to measure indirect emissions. These are the CO_2e that has been emitted as our goods and services are grown, developed, manufactured, transported, used and disposed of. The process of calculating what these emissions are is known as working out the carbon footprint.

The carbon footprint

A carbon footprint is the measure of the emissions released during production, transportation and disposal of a product or service, used by a person, a company, or a nation. The concept of a carbon footprint can be used to measure everything from the emissions associated with an individual's plane flight, to the emissions for which a product, industry sector, or even a country, is responsible for. One of the greatest difficulties is that the boundaries of what should be measured have still not been agreed to a global standard. Most accede that it should also cover direct emissions from the burning of fossil fuels, including domestic energy consumption and transportation. Others believe it should include direct and indirect emissions, generated from the complete lifecycle of products and services we use.

'Nearly everything we use, wear, eat, even clean with contains embedded carbon and that can be very difficult to track.'

If a carbon footprint includes the emissions expressed in the production of goods and services, and those involved in their consumption, it helps us to account for these emissions, no matter where they were actually generated. This can help us identify where responsibility for emissions lies. At the same time, a carbon footprint analysis helps identify the activities that have the greatest emissions associated with it, and can be used as the first step in emissions management. It is also the most effective available means of reporting and managing carbon emissions. That does not mean that developing a carbon footprint doesn't have problems however, as there are inconsistencies between a wide range of reporting protocols as well as a lack of consensus on how and where to measure emissions.

Carbon footprint methodologies and confusion: why is it so difficult?

One of the barriers to effective carbon footprint measurement is the amount of work involved in creating a sufficiently rigorous analysis. There's no doubt that direct emissions, from energy used in buildings, manufacturing and transportation, and indirect emissions, through the use of goods and services, can be calculated. But at an individual product level this can be very difficult to do, particularly because of the difficulty in estimating embodied or embedded carbon. These are terms for the total amount of CO_2 emitted during the production, manufacture and distribution of a given product. Nearly everything we use, wear, eat, even clean with contains embedded carbon and that can be very difficult to track.

The first big attempt to measure and cut GHG emissions came with the introduction of the EU's Emission Trading Scheme in 2005. The process seemed simple as it was focused on the emissions generated through the combustion of fuel, and the measurable emissions generated from specific industrial processes. This attempt required the measure of CO_2e generated in chemical reactions in five key areas associated with large emissions: power generation, oil refining, mineral industries (such as cement, glass and ceramics), steel and other metals, pulp and paper and other large combustion plants.

Even so, effective analysis of the information remained difficult. Emissions reports were required from individual installations rather than companies. Each national registry reported emissions at an individual installation, not company-wide, level, which can make it difficult to understand. Many of the regulated companies have installations in different countries, and were reporting different emissions measurements, which made it even harder to find accessible information.

Things were even more complicated outside the regulated market, due to the range of footprint methodologies used by companies to calculate their footprints. A 2008 report from Ethical Corporation on emissions reporting showed that 34 different protocols and guidelines were used by FTSE 500 companies in responding to a request for information from the Carbon Disclosure Project (CDP).

The range of methodologies and protocols used to report emissions and strategic emission reduction goals makes it hard to make any sense of the information submitted. There are a broad range of reporting methodologies, from carbon emissions per kWh, to the carbon emitted over the lifetime of an event, or over a year (as in many of the installations reporting). Reporting can cover CO_2 emissions, or CO_2e, which covers all GHGs emitted through business. The information can then be used for a report on absolute emissions levels; or emissions intensity e.g. relative to output. This makes tracking what companies are actually doing extremely difficult. And if we can't get clear information from companies, how are we supposed to be able to calculate the footprint of the products and services that they provide.

Where should we measure carbon?

The most commonly used guideline today is the Greenhouse Gas Protocol, a widely used standard developed by the World Business Council for Sustainable Development and the World Resources Institute. This splits carbon emissions' reporting into three areas:

Scope 1	Direct emissions from activities controlled by the organization (from fuel use in the workplace, gas for hot water)
Scope 2	Emissions from the use of electricity
Scope 3	Indirect emissions from the upstream and downstream footprint of products and services (eg. the emissions involved in the preparation and transportation of raw materials, or the use and disposal of goods).

Part of the problem is where we define boundaries. What happens if an emission is counted twice, or not at all? If there are 10 tonnes of carbon associated with transporting goods to your store, do you include that emission in your company's footprint? Or does it belong to the transportation company?

The first step towards a global standard

At present, there is no mechanism in place for analysing the overall carbon footprint of the majority of goods that are manufactured. At the consumer level, there are over 300 different ecolabels on the market, ranging from Fairtrade to food miles to organic accreditation and it can be confusing. In 2008, however, the UK launched a carbon labelling standard, following a two year project by the Carbon Trust, Defra and the British Standards Institute to develop a methodology for measuring carbon in individual products.

The UK's Publicly Available Specification, or PAS 2050, will supply information to manufacturers on how to calculate a specific product's carbon output, from raw materials, through manufacturing, distribution and consumption, even through to the methods of waste disposal used. The idea is to use that information to label goods with their carbon content, thus informing consumers.

While the scheme is voluntary, such companies as Coca–Cola, Cadbury, Tesco, Walkers, Boots and Innocent are also working with The Carbon Trust. The Carbon Trust has even opened an office in China in partnership with the China Energy Conservation Investment Corporation (CECIC) in order to develop a full carbon footprint for Chinese products. The project will initially cover measurement of the embedded carbon footprint of 10 Chinese-manufactured products using the PAS 2050 standard.

Other standards have been, or are being developed. There is the Confederation of European Paper Industries (CEPI) framework, which was launched in the hopes of providing a framework for international agreements on carbon footprints. It is based on an analysis of carbon sequestration in forests, the carbon in forest-based products, emissions from forest production manufacturing facilities; emissions from the production of fibre (for virgin fibre, this includes forest management and harvesting and for recovered fibre, it includes collection, sorting and processing of recovered paper before it enters the recycling process); emissions from other raw materials; emissions from purchased power and transportation; emissions from product use, disposal; and, finally, avoided emissions and offsets.

The International Standards Organization (ISO) has a carbon standard in the pipeline – the proposed ISO 14067 (for products). This new standard is currently being developed by international technical groups working concurrently on two parts: Quantification (Part 1) and Communication (Part 2). ISO 14067 is due for completion in 2011. Each of these approaches will be useful but the range of standards under development simply highlights the lack of a global standard. A continuing lack of agreed international benchmarks by sector, as well as poor corporate performance in analysying emissions is constraining carbon measurement.

Things are beginning to change, however. Australia, Canada and the UK (following the introduction of the Carbon Reduction Commitment, a new mandatory cap-and-trade scheme for large non-energy intensive industries), are the only countries where it is mandatory for companies with emissions over a certain level to report on those emissions, but many more are likely to follow. And that means that the need for an international standard is only going to increase. Until a global standard is reached however, the Greenhouse Gas Protocol provides a useful benchmark.

Once we've worked out how and where to measure emissions, we can begin to take a closer look at what technologies and approaches are available to help us cut those emissions.

Decarbonizing Our World

4

Once we have a mechanism for putting a price on carbon, we need to work out which alternative technologies and practices are available at a reasonable price. A suitable carbon price helps change corporate and individual behaviour by encouraging the use of low carbon alternatives. We need to find the fastest, most effective and economic ways to cut emissions, in a way that supports economic development in a fair and reasonable way.

We need to explore all the options that could influence the amount of carbon emitted in the day-to-day operations of the modern economy and we need to find ways of making changes from today. The longer we wait to address rising emissions, the more dramatic the transition will need to be, in terms of cost, ease of adaptation and economic stability.

Many argue that the climate crisis is such that only 'silver bullet' solutions can make a sufficient difference. These range from a transformation of the economy to a hydrogen basis (using fuel cells for the generation of energy in transportation, power generation both industrially and in the home); through the wide-spread implementation of carbon capture and storage; even going as far as to propose geo-engineering solutions. These are planetary engineering solutions that would affect the climate as a whole, including projects to seed artifical clouds to reflect sunlight into space or dumping iron filings in the ocean, to encourage the growth of CO_2 absorbing plankton. While there is growing support for such approaches, the reality is that structuring and funding such global projects is likely to take a long time – time that we may not have if we need to reach peak emissions in the next few years.

The reality is that we already have the technologies available to completely transform the existing carbon environment. The barriers are more about overcoming establishment inertia and a willingness to transform our operational frameworks rather than a technical or financial inability to address the problem.

Stabilizing emissions

The key to stabilising emissions is to take a range of approaches and work on them in unison, not wait for one particular solution in the hopes that it will have sufficient impact.

We could transform power generation by shifting electricity generation to a renewable energy basis and applying efficiency standards to appliances and machinery to lower both residential and industrial power use. Transportation emissions could be cut through improving fuel efficiency, increasing mass transportation and the development of electric vehicles. It's also important that we look beyond purely technological solutions. It's increasingly clear that unless we address the issue of land-use change we run the risk of destroying the ecosystem's ability to absorb the CO_2 already in the atmosphere. If we address the problem in a number of ways, we arefar more likely to reach our goal of 80% cuts in emissions by 2050.

In 2004, Robert Socolow and Steve Pacala outlined the concept of stablization wedges – each wedge representing a set amount of emissions that could be removed by different technologies. Their idea was that by taking a portfolio approach, we can make significant inroads into emissions. Socolow and Pacala identified 15 different strategies that could reduce emissions by 25 billion tonnes over a 50 year period.

This approach means that we can use technologies either already available or at pilot stage, such as wind and solar power or, for the purposes of their paper, carbon capture and storage. The technologies can be implemented over a prolonged period and enable reductions to grow at a certain rate every year. These emissions reductions would be cumulative, which is critical as CO_2 stays in the atmosphere for around one hundred years.

PART 1: DECARBONIZING POWER

The most critical area in need of reform is the power sector, with the generation of electricity responsible for over 40% of global emissions. We're right on the cusp of a new investment cycle. The pan-European electricity lobby group, Eurelectric, has said that the EU alone will need about 520GW of new power capacity by 2030. The EU has admitted it expects 450GW of power to go offline by 2015. This leaves an enormous energy supply gap. The European Commission estimates that in order to keep up with power demand growth EU member states will need to invest over €750 billion in infrastructure over the next three decades, divided equally between generation and networks.

That means that if we act now, we could transform the global carbon environment. If we wait, it could be another 30 years before the next investment cycle comes around. It's certainly likely to prove a lot harder to persuade politicians and power companies to switch from coal-fired power in a few years time if they've just completed billions of investment in new plants.

According to the International Energy Agency's (IEA) World Energy Report, 2008, 'It is not an exaggeration to claim that the future of human prosperity depends on how successfully we tackle the two central energy challenges facing us today: securing the supply of reliable and affordable energy; and effecting a rapid transformation to a low-carbon, efficient and environmentally benign system of energy supply.' It has warned that without a global $35 trillion energy technology revolution, the world could face a 130% surge in carbon emissions by 2050.

The IEA's latest figures say that 60% of all GHG emissions could stem from energy production, which makes taking carbon out of the power sector a critical step. It also projects a year-on-year growth in power demand of roughly 1.6%. This is a dramatic fall from the 2007 prediction of 3.3% growth, due to the combined impact of the economic slowdown, prospects for higher energy prices and some new policy initiatives. Statistics vary according to the economic environment, but the point is that demand is only going in one direction and that's up.

Of course, demand growth will vary in different parts of the world. Energy consumption is growing rapidly in such countries as Brazil, Russia, India and China – up to three times faster than in OECD countries such as the UK and US.

What makes transforming power generation so appealing is that there are so many different approaches, ranging from a change in fuel source, and increased demand efficiency, to transformation of the power distribution network (cutting transmission losses, increasing efficiency and securing local supply). While there are going to be costs associated with that transformation, there are also going to be significant opportunities.

The power sector needs to focus on a technology-based strategy for reducing GHG emissions, one that creates opportunities for economic growth. It needs to utilize a diverse portfolio of advanced technologies, including the enhancement and expansion of power distribution and transmission capacities. Rather than simple replacement of existing infrastructure, now is the time to take the opportunity for fresh thinking and innovation. When looking at how best we should manage this transformation, we must explore alternative fuels, transmission and distribution, the potential of transforming the electricity grid into a 'Super' or 'Smart' grid through the use of new technologies (see pages 108–115), as well as the opportunity provided through the decentralization of power generation.

A wide range of proposals have been put forward but most include increasing amounts of renewable power, increases in energy efficiency, as well as a transformation of the electricity grid. Importantly, some of them attempt to show how savings in energy cost, and opportunities provided by the development of new markets can combine to provide a significant economic boost.

Clean energy 2030

In October 2008, Google announced Clean Energy 2030, a $4.4 trillion plan to make the US completely fossil-fuel free by that year. The plan could potentially result in cutting both oil use for cars by 44% and total CO_2 emissions by 95%. Google's Jeffrey Greenblatt, author of the report, said: 'Google's proposal will benefit the US by increasing energy security, protecting the environment, creating new jobs, and helping to create the conditions for long-term prosperity.'

The proposal calls for increased renewable energy, newly efficient power grids and the acceleration of production of electric vehicles. Google claims that its proposal would return net savings of between $820 billion and $1 trillion over the period to 2030. It would result in the replacement of all coal and oil electricity generation with renewable electricity. This would see a 20-fold increase in wind power to 380GW, an increase in solar power from 1GW to 250GW, and a further 80GW of generation contributed by geothermal power.

Energy [R]evolution: a sustainable world energy outlook

Another proposal was the October 2008 report from the European Renewable Energy Council (EREC) and Greenpeace. It said that aggressive investment in renewable power generation and energy efficiency could create an annual $360 billion industry, provide half of the world's electricity and slash over $18 trillion in future fuel costs. The report also estimates that without this change, electricity supply costs are likely to rise from an annual figure of $1.75 trillion to $3.8 trillion by 2020. If the transformation of the energy environment is achieved, savings of $18.7 trillion – about $750 billion a year – could be achieved by 2030.

The report even proposes a way to make fossil-fuel obsolete by 2090, through a massive shift from private vehicles to public transportation and the replacement of centralized fossil-fuel powered generators with decentralized renewable energy generators. *Energy [R]evolution* is said to be a practical blueprint for ensuring that GHGs peak and fall by 2015, through strict efficiency standards and a widespread increase in the use of renewables in the heating and transport sectors. The report claims that carbon emissions could be cut by 50% by 2050 if its recommendations are followed.

Decarbonizing the fuel source

There are a range of low carbon fuel sources available to replace hydrocarbon-based power generation, but each has its critics. When exploring different low carbon power sources, however, no solution is perfect. Whether concerned about the environmental impact of a wind power station, a tidal barrage or a nuclear power station, there is a dichotomy between the lobbying of environmental groups to increase sources of clean energy and the potentially negative impact of those developments. Environmentalists are torn – they want clean energy but they want to avoid any negative environmental impact. It is possible that as the need to diminish GHGs released into the atmosphere becomes increasingly critical, a choice may need to be made as to whether the emphasis is put on the protection of the local or the global environment.

One of the best ways to lower CO_2 output is to increase generation efficiency through the use of combined heat and power (CHP), which increases efficiency dramatically, or to cut fossil-fuel use in existing power plants by co-firing with non-fossil-based fuel. While there is an important role for these options, one of the first things we need to do is understand how far we can actually meet our generation needs outside the fossil-fuel environment.

Nuclear power has been the bane of environmentalists for the last 40 years but is now being championed as the ultimate in low carbon generation, while new technologies to remove CO_2 from coal-fired power stations are seen as making coal 'clean'. Renewable sources, while low carbon and sustainable, are said to be too intermittent – wind turbines only generate power when the wind blows, solar when the sun shines, etc, meaning that it can't provide sufficient reliable power to meet demand. The cost of using each fuel source is also critical, in order to be able to compete with traditional fossil-fuels.

Nuclear: the ultimate in low carbon power?

Nuclear has come back up the agenda as it provides continuous generation of low carbon power, which is seen as increasingly important in a world where fuel supply is a national security issue. But should we be focussing on nuclear to meet carbon targets? The fuel required for nuclear fission is going to become increasingly in demand if we do, and it's yet to be proven that this is an economic option.

There is an immediate issue to address – the need to replace existing nuclear generation as it ages, either with new nuclear, fossil-fuel power stations, or other alternatives. For example, in the UK in 2007, nuclear power supplied roughly 18% of the country's electricity but 18 of its 19 reactors will reach the end of their service life by 2023. Despite Labour government support for new nuclear, replacing current reactors over the next 15 years won't take the UK much closer to its EU goal of cutting emissions by 15% by 2020. A report from the Sustainable Development Commission (SDC) concluded that replacing all the existing nuclear capacity with new nuclear plants might save 7 million tonnes of carbon by the late 2020s – that is equivalent to only 4% of total UK emissions.

In 2002, the Bush administration launched its Nuclear 2010 programme, with the goal of deploying new nuclear reactors in the US by 2010, but this proved unrealistic. Wall Street has proved loathe to provide private funding for such developments and estimated costs for building a new nuclear plant in the US have doubled from $6 to $12 billion.

The refocus on nuclear can be seen in many countries traditionally opposed to its deployment. In 2007, Sweden overturned a 30-year ban on nuclear development and, in 2009, stated explicitly that nuclear plants could be constructed to replace existing plants nearing the end of their lives. In 2008, Italy announced plans to resume nuclear construction within 5 years, after a 20-year ban, and many experts expect the Netherlands to follow by 2010. Belgium and Spain remain committed to shutting down current nuclear generation but, while Germany retains plans to phase out its 17 nuclear plants, increasing energy prices are putting pressure on those who oppose nuclear power.

Finance remains a critical issue. To date, no nuclear power station has been built without government subsidy. In the 2008 Nuclear White Paper, the UK government identified a cost of €36* per tonne of carbon as the price at which nuclear power becomes economically competitive. Not only is the carbon price significantly lower than this (falling as far as €8 in February 2009) but the collapse in the price of oil in 2008, combined with low demand on emissions credits due to lower industrial activity, means that the carbon price is likely to remain low until the recession eases. This makes the economics of nuclear even more uncertain.

The capital required to build a nuclear power plant is high, construction periods are long and pay-back can only commence once the plant is fully commissioned. The International Atomic Energy Agency (IAEA) states that 35 reactors are in construction around the world, but many have no official grid connection dates, or in some cases, even start dates.

Nuclear waste disposal is another critical issue. The UK government is proposing that this could be managed through geological sequestration. In the US, the plan was to bury nuclear waste in Yucca Mountain – not nice for the concerned citizens of Nevada, who have consistently forced delays in the project. Meanwhile, nuclear waste is stored in around 100 'temporary' facilities around the US.

---

* the carbon price is traded in euros within the ETS.

Safety is a paramount issue for politicians and consumers alike, but the key issue for those considering nuclear as a low carbon solution is the nature of its generation. There is no question that it does provide low carbon baseload power, but output levels cannot be easily changed, which means that it can't respond to peaking demand. In a country like Sweden, with nearly 50% of its power supplied by hydropower, that's not a problem. Hydropower can be turned up and down according to demand.

If, however, we implement large amounts of nuclear power in markets without a significant means of renewable dispatchable power, we have a problem. Any sizeable deployment of nuclear power stations will require an equally substantial number of power stations where the power can be dialled up or down as needed. That predominantly means more gas or coal-fired plants. It's possible that large scale deployment of nuclear power could actually act as a brake on the deployment of other low carbon alternatives. Nuclear's role in a low carbon future then only becomes acceptable hand in hand with a transformation of the nature of the generation, transmission and distribution system in order to avoid the need for increased amounts of coal and gas. Of course, if we could generate clean coal power, then a wider deployment of nuclear seems more reasonable. So can we?

Clean coal: is there any such thing?

'Clean coal' is the idea that coal-fired power stations can continue to operate with the implementation of technologies that take harmful GHGs out of the process. This is known as carbon capture and storage or sequestration (CCS) and would result in emissions being stored in geological formations or old oil fields. This is of great appeal to the oil industry, as gas is already pumped into oil fields to help extract oil, and to the coal industry, for obvious reasons.

Politicians and policymakers seem to accept that coal is going to be part of the global energy mix for the foreseeable future. This is supposed to be acceptable in a low carbon future because of the advent of clean coal technologies, ranging from the chemical washing of coal to gasification, but the central tenet of the clean coal market is the use of CCS technologies.

The IEA World Energy Outlook 2008 warned that current trends in energy supply and consumption are 'patently unsustainable' and that the development of CCS is a critical requirement for any attempt to mitigate carbon emissions. It hopes to see at least 20 projects announced by 2010, with 10,000 CCS projects in place by 2050 and has warned that the cost of global mitigation could increase by 70% without CCS technologies. The difficulty here is that the IEA is predicting the cost and carbon mitigation capacity of a technology that is not yet commercially available, and where arguments still run about the extent of its contribution to emissions mitigation.

'Just because coal is cleaner than it used to be that doesn't mean that it has lower CO_2 emissions.'

But is 'clean coal' actually clean? And is it economic? If we concentrate on modern coal-fired plants, there's no question that most coal is far cleaner than it used to be – a lot of work has been done to cut the emission of pollutants, such as sulphur dioxide, driven by legislation to cut acid rain levels. However, just because it is cleaner than it used to be that doesn't mean that it has lower CO_2 emissions.

Some commentators suggest that viable large-scale CCS technologies could eventually capture about 90% of the CO_2 emitted by current fossil-fuel-powered plants. A number of CCS pilot plants are already in development in Europe, the US and Australia. But coal emits more CO_2 for every unit of energy generated than any other fuel and to call coal 'clean' because of the potential to capture and store emitted CO_2 at some point in the future seems misleading, at best. At present, there are four full-scale CCS projects in place, none of which have been capturing CO_2 from coal-fired power plants at the beginning of 2009.

Within the EU, the mandatory CCS is being debated in terms of its role in post-2020 regulation. With about 50 coal-fired power plants due to come online in Europe by 2013, the debate is critical. The question of whether a focus on CCS is the best way to approach reductions remains open. But surely the focus should be on technologies available to cut emissions today?

With the technology at the very least 10 years away from commercialization, however, we need to think very carefully before putting all our eggs in one basket. Massachusetts Institute of Technology (MIT) published a 2007 report called The Future of Coal, in which it argued that the first commercial CCS plant might not appear until 2020. That was before the US decided to cut funding for its key $1 billion FutureGen demonstration plant, apparently due to rising costs.

Utility-scale deployment (meaning large scale, industry-wide deployment) is not really expected before 2030. On the surface, the position in the US looks slightly different. Following a 2007 US Supreme Court ruling that CO_2 could be regulated as a pollutant under the 1970 Clean Air Act, environmental groups have been pushing the EPA to stop issuing permits to coal plants. Yet the US is heavily dependent on coal for power generation, and many states have significant coal interests. Over 90% of coal in the US is used for electricity generation, resulting in 83% of CO_2 emissions from the power sector and CCS has been positioned at the heart of the US drive towards energy security. In March 2009, US Energy Secretary Steven Chu said that the Obama administration was moving forward on loan guarantees for nuclear power and how to allocate funds for clean coal projects – potentially involving FutureGen.

Australia, one of the world's largest coal exporters, is home to 13 of the 40 or so retrofit or carbon storage trials going on around the world, funded by the Coal21 Fund – a AUS$1 billion fund (raised from a voluntary levy on the coal industry over the coming decade) dedicated to fast-tracking the demonstration of low emission coal technologies.

In New South Wales, CSIRO is leading a post-combustion capture trial and the same team is working with China's largest power producer, the Huaneng Group, on developing a joint pilot to recover 85% of CO_2 from power station flue gases. Given the huge growth in coal power in China, this is potentially exciting, but it's not available now. In the developing world, the idea of CCS remains a pipedream, with hundreds of megawatts of dirty coal being added to the power generation network every week in China alone – a country dependent on coal for about two-thirds of its energy use.

This has been a major driving force behind China's rise in the ranks of major global emitters of GHGs. The True Cost of Coal, a report from Greenpeace China, stated that the country's heavy reliance on coal-fired energy costs the country about $250 billion a year or 7% cent of China's GDP in 2007, including the hidden cost of coal in human deaths, poor health impacts and environmental damage.

The outlook for the next 10 to 20 years doesn't look much better and there's one very simple reason for this – planned coal-fired power stations have no requirement for CCS facilities. The majority of coal-fired plants in planning in the developed world are simply being asked to be 'CCS-ready' but clarity about what that actually means is hard to find. The plants being built now could be in operation in 30 years and with no clarity on the technology, or price in sight, could simply continue operating business as usual, with nothing clean about them at all.

As CCS is still in the early stages of development, it is estimated that the technology will add $1 billion to the cost of a large-scale coal-fired power plant. CCS has other problems, in that capturing and compressing CO_2 is likely to decrease efficiency between 10 and 40% (dependent on technology) and could use as much as 40% of the energy generated by a power plant. This almost doubles the price of the electricity delivered to the power grid. A 2008 McKinsey study estimates that adding clean coal technology could increase the capital cost of a power plant by 50% – and that doesn't include the costs of transporting and storing the captured CO_2 or operating costs.

In 2009, the UK government took the first step in clarifying what CCS ready might mean, announcing that all new UK coal-fired power stations must have demonstration CCS capabilities on at least 400MW of output, about a quarter of the output of an average power station, from the first day of operation. New plants would also have to demonstrate that they will fit CCS systems capable of capturing 100% of their carbon emissions by 2025. What remains to be seen is how this will be funded, and how effective it will prove. At the time of the announcement, Greenpeace calculated that four average sized CCS demonstration plants operating for 15 years before becoming fully CCS, would still emit up to 275 million tonnes of CO_2.

Even if the resulting CO_2 was captured and stored, the associated increase in coal consumption would cause continuing environmental degradation. The environmental consequences involved in the production of coal, from both strip and underground mining, can be harsh. The process of extracting coal can result in soil erosion, groundwater contamination, habitat destruction and toxic waste.

It is not just technological viability or unavoided consequences that make CCS a questionable choice. There is also strong disagreement within the coal industry about the price point at which CCS becomes viable. Swedish power company Vattenfall has said that the technology becomes economic at a carbon credit price of €35–40 per tonne and that, as such, it has the capacity to transform the coal power market. This would make CCS a very reasonable option under the Vattenfall Abatement Curve (see page 57). But some experts believe that CCS will only be viable at a carbon price of €90 per tonne or more. With the economics uncertain, the technology for carbon capture unproven and the storage potential for liquified CO_2 unclear, while there may be potential in the CCS market, it is clear that we can't afford to focus all our investment and energy in just one direction.

Renewable alternatives

Renewable energy is critically important in the decarbonization of power generation, for three key reasons:

1   if a robust renewable energy framework is created, it can play a major role in securing energy independence.
2   as the cost of oil, coal and natural gas increases the use of a renewable fuel source can offset those increasing energy costs.
3   the GHG emissions associated with the operation of a renewable power plant make it a viable long-term option for a low carbon future.

One of the key issues for developing alternative energy technologies is the cost. Constructing a coal-fired power station remains far cheaper than developing a similar sized project in solar or wind, for example. Once we look at capital cost versus fuel cost, however, the economics begin to change. Development of renewable power plants has an economic advantage over the long-term – once the plant has been developed there are few further costs. Given that the long-term cost trends for renewable energy are coming down, while those for hydrocarbon fuel are going up, integrating renewable alternatives into the fuel mix is likely to have benefits in terms of CO_2, cost and security of power supply.

Different renewable fuel sources work more effectively in different markets, either in terms of scale, geography or ease of deployment. Wind, solar, tidal and wave provide intermittent power at a range of scales; geothermal and hydropower can provide large-scale base load power but are limited by geography; while new technologies such as fuel cells are still on the path to commercialization.

Wind power

In terms of cost and deployment, wind power has proved the most sucessful of renewable fuel alternatives, driven by a combination of policy and fiscal support. According to the Global Wind Energy

Association, global wind energy capacity grew by 28.8% in 2008 to reach total global installations of more than 120.8GW by the end of the year. That means that over 27GW of new wind power generation capacity came online in 2008, 36% more than in 2007. The question is what role wind can play in transforming the energy environment?

There is enough potential wind power to generate more electricity than the entire world uses today. A 2005 study by Stanford University's Global Climate and Energy Project, found that wind could generate more than 7 times global power consumption, even if only 20% of it was accessible for actually building wind power plants. Some of the strongest winds were observed in Northern Europe, along the North Sea, while the southern tip of South America and the Australian island of Tasmania also featured sustained strong winds. North America had the greatest wind-power potential, however, particularly in the Great Lakes region and from ocean breezes along coasts.

The study found that the locations with sustainable Class 3 winds, the most suitable for generation, could support installed capacity of approximately 72 terawatts or TW (one terawatt equals 1 trillion watts, the power generated by more than 500 nuclear reactors or thousands of coal-burning plants running at full capacity). Capturing even a fraction of those 72TW in actual generation capacity could provide the 1.6 to 1.8 terawatt hours (TWh) that made up the world's electricity use in 2000. Converting as little as 20% of potential wind energy around the world to electricity could satisfy all of global energy demand. Of course, there are regions where that power is more accessible but the study clearly shows that wind provides a viable option.

There are benefits to using wind. It's cheaper than solar and its utility scale and efficiency are high. But traditionally wind power plants have cost more to build than coal power plants, at an average cost of around $1–2 million per MW for a fully installed wind farm (including roads, power lines and development costs). At the same time where winds are most accessible are often far distant from the point of use.

Wind plants have often also been the target for environmental complaints. This is usually an issue for major large-scale wind farms, both on and offfshore. However, wind can also be used as a power source on a distributed generation basis, with small wind farms providing local power to manufacturing or agriculture, even to local mini-grids. There has been promotion of the idea of wind turbines on individual homes, but deployment is likely to take somewhat longer, given the need to implement other technologies to integrate individual turbines into the local network, net-metering to enable local utilities to know what power is being generated and so on.

While the question of cost remains important, the major criticism of wind power is its variability. Opponents of renewable energy have said that the intermittent way that the wind blows means that it can never be used as a grid level power source. Over the course of a year, a wind turbine will usually generate about 20% to 30% of the amount it would generate in a constant strong wind – its load factor (or capacity factor). The turbine itself has usually at least 97% mechanical availability and generates power with 95% mechanical to electrical efficiency, so the limited output is overwhelmingly associated with lack of wind. It's worth noting that the load factor of conventional thermal power stations (such as coal-fired power plants in the UK) is an average of 50%.

Although wind may be intermittment that doesn't mean it's unpredictable – knowing when the grid is going to receive power is the first step to using it in a balanced way. Accurately forecasting when the wind is going to blow is a vital aspect of operating a wind farm providing power to the grid. Wind energy is already predictable enough to make seasonal commitments for the displacement of natural gas and, in short-term forecasting, wind fluctuations in early morning or evening can even match spikes in power demand.

The power industry has a long history in forecasting electricity load and future electricity spot prices – many are now using resource forecasting tools built for such projections to forecast wind. According to the Institute for Energy and Environmental Research (IEER), costs for wind integration into the grid can be greatly reduced by increasing wind forecasting accuracy and standby capacity can be reduced by about half at high penetration levels, for a 40% reduction in forecasting error.

Solar power

Solar could solve our energy requirements many times over. The amount of solar energy that reaches the Earth's surface every 20 days exceeds the energy trapped in all of the planet's coal, oil and natural gas reserves. The trick is finding cost-effective and efficient ways of converting this abundant resource into usable energy.

Standard photovoltaic (PV) cells have an energy conversion rate (the amount of solar power that falls onto the panel that is turned into energy) of 6–18%, with crystalline silicon typically at 13–16% and current flexible modules are around 6%. Some prototypes have already achieved conversion rates of more than 40%, but are still too expensive for mass-market production.

Solar PV is well-known for having lower conversion efficiencies but there are plus points. A solar installation is less likely to create problems with local communities in terms of planning or concerns about the impact on wildlife and the local environment. Load balancing of solar micro-electricity on a distributed scale (in residential and office installations) could also negate the need for utility scale grid load management or increased energy storage. And of course, the efficiency isn't that important for solar – the input is effectively limitless.

Growth potential is a key driver for the solar market – there are improvements in technology and efficiency, creating huge potential for costs to fall. With increases in efficiency and large amounts of solar manufacturing coming online, the price will continue to fall. According to investment bank Merrill Lynch, solar is on the steepest price–performance improvement curve of all renewable technologies.

One new development is thin-film solar, which is faster and cheaper to manufacture than traditional PV, as well as having the advantage of working more effectively under indirect lighting conditions, such as in cloudy weather. The 2007 edition of Greentech Media's Thin-film PV 2.0: Market Outlook through 2012, estimates that thin-film could make up over 40% of worldwide PV production by 2012 and suggests that it will be providing 10GW of production capacity by that time.

Another growth opportunity lies in integrating solar generating capacity into roof tiles, windows and more. This will give developers the ability to directly incorporate solar PV into buildings as passive electricity generators – multifunctional building panels. Australia's Dyesol, which has developed a dye-based solar cell technology, is even working on developing its technology on steel-sheeting building products, in partnership with steel producer Corus.

One of the most exciting options is the opportunity provided by solar-thermal generation, also known as 'concentrated solar power' (CSP), which uses the Sun to heat a working fluid to drive a steam turbine and generate power. Much of the interest lies in the ability of CSP to function in utility scale plants.

CSP or large-scale solar-thermal-energy generation (STEG) is already being explored in the Middle East, Africa, Europe and the US. As the plants often also use hot steam to power turbines, it's not so different from the gas- and coal-fired plants that utilities are already familiar with, at an efficiency that can range from 23 to 31%. The need for large amounts of space is key, as the collectors must cover a wide range. Another critical issue is water – CSP plants can require hundreds of millions of gallons to keep the collectors clean. This can be a problem in regions best-suited to CSP in terms of insolation and geography, where it is typically very dry. There are interesting ideas being developed however, regarding desalination for plants in coastal areas, such as the Seawater Greenhouse.

CSP has a useful storage side benefit as well, with the heat generated through the process warming salts. These molten salts are capable of storing excess thermal energy meaning that power generation can be maintained even after the sun has gone down.

While nine CSP plants were deployed in the US between 1984 and 1991, with a combined capacity of 354MW, progress ground to a halt after natural gas prices fell in the 1990s. It wasn't until 2007 that the next major plant in the US opened: a 64MW parabolic trough system in Nevada, built by the Spanish company Acciona. Now there are at least 13 other plants, totalling 5100 MW, in planning stages in Florida, Arizona and California; most will use parabolic troughs although there are a range of alternative technologies. Israel is also exploring the possibility of developing a 500MW plant, while a number of projects are in development in Europe.

In the short-term it is likely that PV will continue to outstrip STEG development, as it takes around three years to develop a STEG plant. Price will also remain an important factor in its take-up: in the US, coal-fired plants generate at 5–7 cents per kWh and US-based STEG plant Nevada Solar 1 is selling power at 18 cents per kWh – so it must be generating power at a cost below that level. The differential exists but it is continuing to fall.

By providing utility-scale power and by providing distributed power to take pressure off the existing grid, solar technologies can significantly contribute to a cut in demand for fossil-fuelled power.

One thing we shouldn't forget, of course, is the traditional form of solar thermal on the home front, which is using solar heating units to heat water and cut power demand. By far the largest solar thermal market in the world, according to newly installed solar thermal capacity per year, is China. In 2007, around 15.4 GWth (22 million m2) was sold in China, which provided 77 % of the world global solar-thermal market with a total of estimated 19.8 GWth. In Europe, Germany – the second biggest market in the world – remains dominant. With its newly installed capacity of 658 MWth (940,000 m2) in 2007, the country reached a market share of 34 %.

Hydropower

This is the largest contributor to renewable power, providing around 20% of global electricity generation and around 2% of the global energy supply. This is hardly surprising as about 70% of the Earth is covered in water. The WWF estimates that there are resources of about 370GW of hydro-power which could be used for power generation without 'unacceptable impacts' by 2050.

Hydropower can provide both baseload or despatchable power, which makes it a vital part of the power generation mix. The fact that reservoirs can also be used to store power (using power to push water back up to the reservoir and releasing it to power turbines as and when it's needed) gives it a compelling role in ensuring power security in mixed power systems. Another driver for hydro development is the increasing need for water management as multi-purpose hydro reservoirs can bring security of water supply as well as power.

One of the downsides for major hydropower developments are the environmental issues. Major dams, for example, forcing people off their land, drowned ecosystems and influence on neighbouring water systems to name but a few. However, a widely distributed network of small scale hydropower systems could well be of an advantage in a decentralized power network.

There is a concern regarding hydro power, however, and that's the potential impact of drought conditions. In 2006 in Spain, dams for hydroelectric production and irrigation fell to about 40% of their capacity. Hydroelectric power generation fell to its lowest in 48 years during the drought of 2005. This can have a direct knock-on effect on the need for alternative power generation and storage.

Wave and tidal power

The World Energy Council has said that if we could harness just 0.2% of wave power we could provide enough power to meet global demand. Globally, about 10% of current power demand could be generated via existing wave technologies but the entire sector remains at a relatively early stage of . The economically recoverable resource for the UK alone has been estimated to be 25% of current demand, while Australia is estimated to receive 1m GWh of potential wave energy annually. The technologies exist and, while they may not yet be fully commercial, we can certainly expect some contribution from wave power by about 2030. There's even the possibility of exploring ocean thermal-energy conversion, which would generate energy by using the differences between surface and deep-sea ocean temperatures, particularly in tropical climates.

Tidal power, while also in the early stage of commercialization, seems further along than wave power. The sector is expected to grow dramatically over the next 20 years but that means that its deployment should be encouraged, without necessarily being a focus for today's energy solutions. At the same time, there are significant environmental issues raised by the use of tidal power, which can be illustrated by the UK's plan for the Severn Barrage.

The UK is a leader in the tidal sector, and it has been projected that its tidal resources have the potential to supply at least 10% of UK electricity needs. The Severn Estuary is one of the largest in the UK, with the second highest tidal range (at 14m) and the proposed Severn Barrage Tidal Scheme could supply almost 5% of the UK's electricity alone. There is, however, immense environmental opposition to the project.

Conservationists are staunchly opposed, saying that the impact will be devastating on the thousands of birds that winter in the area and 11,000 acres of tidal and protected land. Some, such as Friends of the Earth, suggest that alternative technologies, such as the use of tidal lagoons, could prove more effective in generating power (with the potential of up to 60% more energy than a tidal barrage) without the concomitant damage to the region.

Geothermal power

Geothermal power is the third-largest renewable power source, behind hydroelectricity and biomass, ahead of solar and wind. It is generated using underground steam or hot water to turn turbines to create electricity – heat created by volcanic activity, often at the edges of tectonic plates. Geothermal power has a great deal of appeal as it is independent of weather conditions, has inherent storage capability and can be used both for base-load and peak power plants. Its only real problem is that it is totally dependent on geography and geology.

Electricity is produced from geothermal sources in about 25 countries and 5 of these obtain between 15 and 23% of their national production from geothermal, including the Philippines, Iceland and Indonesia. Exploitable geothermal systems occur in a number of geological environments, from shallow heat which can be used in district heating (cutting fuel requirements) all the way through to high temperature power generation.

While there has been little innovation in the geothermal sector, one technology that has been under development for around 35 years, called Hot Dry Rocks, has been gaining attention. The idea is to create an artificial reservoir, or heat exchange, within the rock, without the need for tectonic activity. In 2008, the online search giant Google invested over $10 million in a company called Enhanced Geothermal Systems (EGS). The EGS process replicates tectonic conditions by fracturing hot rock, circulating water through the system and using the resulting steam to produce electricity in a conventional turbine.

While there is some way to go before Hot Rocks becomes an easily accessible power source, geothermal has more to offer. The use of low and moderate temperature geothermal resources are potentially most interesting for distributed, decentralized local generation – direct use and ground-source heat pumps. Direct use, as the name implies, involves using the heat in the water directly for such things as heating of buildings, industrial processes, greenhouses, aquaculture and resorts, and generally requires temperatures of between 38°C (100.4°F) and 149°C (300.2°F).

Combining such heat with other power generation can fundamentally change the efficiency of generation and local power demand, by 3–4 times. Ground-source heat pumps use the earth or groundwater as a heat source in winter and a heat sink in summer. With temperatures of 4°C (39.2°F) to 38°C (100.4°F), using the heat pump to move heat from one place to another, the system transfers heat from the soil to the house in winter and from the house to the soil in summer. Unlike solar heating, which works worst when you most need it, in the middle of a cold winter, geothermal energy is not affected by the weather. Geothermal heat pumps are one of the fastest-growing technologies available and are gaining wide acceptance in use for both residential and commercial buildings.

Biomass, biogas and waste

Biomass, or organic matter, is the leading renewable energy source in the world, as people burn organic matter such as wood or guano for fuel. According to the World Energy Council, biomass supplies a little over 10% of primary global energy supply. There are a number of ways in which biomass can be used to generate power, from the individual and local, through to large-scale generation of baseload power.

This means that biomass could provide the balance for intermittent sources of generation such as wind and solar. Burning biomass directly has a carbon advantage – although the burning of a tree releases CO_2, it only releases the amount that the plant removed from the atmosphere while growing. Thus, theoretically, as long as the plants that are burned are replaced by growing new vegetation, the net emission of CO_2 is zero.

The main approaches to generating electricity from biomass are:

- direct-firing systems which burn biomass to produce steam to drive turbines.
- co-firing systems which replace parts of the coal burned in existing power plant furnaces with biomass
- gasification which turns biomass into gas through combustion in the absence of air.

There are problems in using biomass for power generations, as directly raising energy crops for combustion has raised a number of issues. Problems with growing fuel crops include seasonality, standardization and the logistical challenge of transferring sufficient feedstock to the power plant. This can mean that biomass power generation is most effective at a local level, using agricultural, wood and municipal waste. Power plants using agricultural waste can provide a significant contribution to power generation, using feedstocks ranging from bagasse (sugarcane residue), rice hulls, rice straw, nut shells, crop residues and prunings from orchards and vineyards.

Co-firing is becoming increasingly popular with coal companies, as it cuts down the cost of fossil-fuel sources. The UK's coal group Drax co-fires in its coal plants: around 3% of their fuel was biomass in 2007–08. In 2008, the company announced plans for a £2 billion programme to build three 300MW biomass-to-energy plants in the UK, alongside ancillary biomass logistics and processing facilities. Drax says that once all the plants are operational, they will be responsible for supplying at least 15% of the UK's renewable power and up to 10% of total UK electricity. This is a significant contribution to baseload power.

Biogasification, the conversion of biomass to biogas (such as cow dung to methane), works at two levels:

- at a household level, the development of biogas stoves is improving health as incineration of wood sources decline.
- biogasification has the ability to produce twice as much energy as most thermal gasification technologies, which are dry processes.

Large scale gasification or anaerobic digestion facilities can provide significant amounts of fuel gas at the local level. Perhaps the most exciting way to look at this opportunity is in terms of the markets it can address. Obviously we are looking for fuel sources which are not fossil-fuel based, but finding ways to transform part of our waste stream into heat and power attacks the problems of both diminishing energy supplies and increasing waste streams. The potential of using our waste as a baseload power source could transform local city and town environments.

To date, municipal solid waste (MSW) or household waste, as well as the waste from buildings and industry, have generated power through the creation of landfill gas. This is where decomposing waste in landfill generates methane, which can be used to power gas turbines. Environmental concerns about pollution from landfills has encouraged the development of such power projects, as well as the reality that methane has 26 times the global warming potential of CO_2.

In the developed world, legislation is encouraging a move away from landfill dumping and towards increased waste recycling. The mantra 'reduce, reuse, recycle' is a clarion call for many. By cutting down on unnecessary waste we can conserve the resources we have but, at the same time, new technologies are potentially offering new ways of generating power from waste.

While incineration plants have become more efficient over the last few years, that remains only relative. Incinerators have large plant footprints, emit a number of pollutants and there are residues following the incineration process. In order to be financially viable, such plants often have to be large. Such scale makes the power generated easier to integrate into existing grid networks but low efficiencies mean that large quantities of waste may need to be trucked in from the surrounding regions, which can increase pollution. And because they're on a large scale, they can't really help support local heat solutions when operating in Combined Heat and Power (CHP) mode, because their size prevents them being located close enough to local users. CHP plants are far more efficient than any normal thermal plant. Combining heat and power generation, whatever the fuel source, increases energy efficiency from around 55% to over 80% efficiency.

What are most exciting are new and alternative waste treatment technologies that can recover products from waste that would be missed by incineration and generate power, but on a far smaller scale. These include composting, pyrolysis (a form of incineration that chemically decomposes organic materials by heat in the absence of oxygen), mechanical biological treatment (MBT), anaerobic digestion, biomass co-firing and gas plasma. Pyrolysis can result in a liquid oil that can be used as fuel, as well as gases to run steam turbines. Gasification can be used with a far broader range of feedstock and simply involves heating wastes in a low-oxygen atmosphere to produce a synthetic gas, which can then be used to power a steam turbine.

Many of the new technologies have multiple advantages over more traditional approaches, both environmentally and in terms of payback. Smaller plants impose lower transportation pressures, and these new technologies rarely result in the leachate and toxic fly ash so frequently associated with incineration. Small scale, local plants, generating power from local waste, enabling the development of baseload power for a local or mini-grid, be that industrial, educational or for a local authority, provide solutions to waste, carbon and energy security concerns. With commercial waste predicted to increase by 50% by 2020, the ability to build small plants in industrial areas could transform the energy landscape with local waste being used to produce local electricity.

On a small industrial, agricultural or municipal basis, there are strong arguments for biomass energy generation, especially where there is a reliable source of waste or energy crops to be used. Anaerobic digesters may offer viable opportunities for small-scale biogas production from animal wastes and are even now being trialled for use in water and sewage plants in the US.

Intermittency and energy storage

One of the great criticisms of renewable power is the intermittency of supply, which is seen as putting increasing pressure on national grids. Traditional grid management requires frequency and voltage, as well as flows of power into and out of the grid, to be equal. So it's clear that as we increase the levels of renewable power in our electricity systems, we need to look at different ways of managing the impact of that intermittency.

As the percentage of energy produced by renewable sources increases, different technical and economic factors affect the need for grid energy storage facilities, demand side management, grid import/export and other management of system loads. Large networks, connected to multiple wind plants at widely separated geographic locations, can accept a higher penetration of wind than small networks or those without storage systems compensating for the variability of wind. In Denmark, Germany and Spain up to 20% of intermittent renewable power has already been added to the grid.

Energy storage should be what enables larger scale integration of intermittent power. It can smooth out spikes in power supply, strengthen the grid by storing power near its point of call and transform intermittent output into a smooth predictable power flow. It could be a crucial hurdle to overcome, for both wind and solar and dramatically improve the economics of renewable generators, making them more appealing as an alternative to fossil-fuel generation.

Much has been made of the possibilities of the 'hydrogen economy' and the use of hydrogen as a portable energy-storage method. If intermittment power sources are used to generate hydrogen, they could be fully utilized when the sun is shining, or the wind is blowing, as hydrogen would simply be stored and used as required. However given the current state of development, this is a high-cost option. The efficiency of hydrogen storage is around 50–60% lower than pumped storage or batteries. And about 50 kWh is required to generate a kilogramme of hydrogen by electrolysis, so the cost of power is a crucial factor. At the same time, the costs of the electrolyser, compressors, storage and transportation all must be taken into consideration.

Another storage method is to use off-peak electricity to compress air, usually stored in an old mine or useful geological feature. When the power is needed, the compressed air is heated and goes through expanders to generate electricity, at efficiencies of between 55–65%. Thermal energy storage is another option, where electricity is used to make ice, which can be stored either for cooling air in a building (shifting the power demand for air-conditioning off-peak) or cooling the air-intake of a gas turbine generator.

Two of the most popular 'new' technologies for power storage are flywheels and superconductors: Flywheels use mechanical inertia, during which a rotating disc is accelerated by an electric motor and the electricity is stored as the kinetic energy of the disc. In order to extend the storage time, friction must be kept to a minimum, often achieved by keeping the flywheel in vacuum and using magnetic bearings. This tends to make flywheels very expensive. Flywheels are most useful as a load-management tool, providing high bursts of power for short durations.

Superconductors store energy in a magnetic field created by a flow of direct current in a cryogenically cooled superconducting coil. The charge can be stored indefinitely and can be released by discharging the coil. The systems are highly efficient (at around 95%) but their high cost has meant that they have not yet become commercial. Aside from the cost of superconductors, the energy requirements of refrigeration and the limits in the total storage capacity have limited their use. They are most commonly used as a means of improving power quality within the grid.

Batteries are the best known devices available for energy storage but the usual options of lead–acid and nickel–cadmium batteries suffer from low storage capacity and relatively high cost. Unfortunately, using batteries for power storage only makes economic sense when the marginal cost of electricity varies more than the energy losses of storing and retrieving it. Of course, the economics can change when you integrate power storage with intermittent generation.

Perhaps the most exciting aspect of the integration of energy storage with intermittent generation is the potential for expanding the power network beyond the grid network. The potential to utilize power generation in conjunction with non-grid connected batteries is interesting, whether in home appliances or even in-the-field electric vehicles. We'll take a closer look at this opportunity when we look at transportation (see page 129).

When it comes to intermittent power, however, it's possible that we don't need to worry about energy storage as much as we think. Given that renewable power, while intermittent, is predictable, perhaps what we really need is to rethink how we manage our power network.

Decarbonizing power infrastructure

Cutting carbon out of power generation means more than increasing the efficiency of generation, transmission and distribution – it means finding new ways to integrate intermittently generated power into our energy mix.

Transmission and distribution

Many OECD countries are at a critical point in their energy investment cycles and the need to replace obsolete equipment provides them with an opportunity to develop smart technologies appropriate for 21st-century challenges. There is little doubt that massive investment is needed, although part of the problem lies in ascertaining who will take fiscal responsibility for such development.

Power blackouts and brownouts (where power supply is temporarily reduced and customers receive a weaker current) are an accepted part of access to electricity in many parts of the developing world. These are due not only to lack of sufficient power plants but also to problems with the grid. The Karachi Electric Supply Company in Pakistan, for example, admitted in 2008 that it was facing power losses of 38–40% due to problems with transmission, distribution and commercial loss (which is where locals simply tap the grid for power without contract).

In the US, a 2008 report from the North American Electric Reliability Corporation (NERC), the industry body responsible for the reliability of the grid, said that the overhauled electric system that has emerged in North America in the last couple of decades already has inadequate transmission capacity to meet demand. It also warned that while there are significant renewable energy sources available to add power to the grid, this would be impossible without investments in transmission.

In North America, the most widespread electricial blackout ever reported occurred in 2003. The Northeast Blackout affected around 10 million people in Ontario, Canada, and around 40 million people in eight US states. It is reported to have cost around $6 billion, through lost production, food spoilage and overtime, as well as indirect costs through secondary effects. Many backup generation systems failed and it wasn't just the lights that went out. Some phone systems failed due to overload, water systems lost pressure in several cities and some TV and radio stations were knocked off air. Hot weather was believed to be a contributing factor as people turned up their air conditioning, causing transmission lines to sag as stronger currents heated the lines, resulting in a cascade failure. This is something that can easily happen and the frequency and magnitude of blackouts and brownouts, once seen as a developing world problem, is also increasingly impacting on network systems in the developed world.

Super Grid

One large-scale solution under discussion is the development of a Super Grid. This is where key electricity markets are connected, enabling large scale movement of power from the point of generation to the point of need. It combines reliable long distance power transmission with the integration of wide-ranging power generation. An international Super Grid could treat wind and solar (and other such renewables) as transnational resources which would enable all participants to share in the enormous energy potential, to their mutual advantage.

The Global Energy Network Institute (GENI) has been exploring Buckminster Fuller's proposal for a global electric energy grid for the last 20 years. GENI research suggests that the best long-term strategy is the interconnection of electric power networks between regions and continents into a global energy grid, with an emphasis on tapping abundant renewable energy resources – a worldwide web of electricity. Their research has underlined the benefits this will bring: linking renewables between all nations could help to dissipate conflicts, grow economies and increase the quality of life and health for those joined to the system.

While there are major issues to overcome in reaching a global agreement on the transfer of power, we are already seeing momentum to connect national grids. The BritNed Interconnector is a 260km long, 1,000MW sub-sea cable that will link the UK and Dutch national grids by 2010. The aim of the €600 million project – a joint initiative between the UK's National Grid and its Dutch equivalent Tennet – is to install a two-way high-voltage direct current (HVDC) sub-sea electricity cable between the two countries.

The cable will be connected into the AC grids of the two countries through two giant converter stations being developed by Siemens. If all goes to plan, the cable should be in operation by the beginning of 2011. Although the UK already has one 2,000MW link to France and another connector linking Scotland and Ireland, BritNed will be the first new large UK interconnector for two decades.

While the key aim behind the BritNed Interconnector is to drive competition in the wholesale electricity market, once those connections are made they can be used to encourage renewable power integration. One of the more exciting projects is a proposal being explored by the EU's Institute for Energy to build a 5,000 km Super Grid stretching from Siberia to Morocco and Egypt to

Iceland, which could integrate the use of renewable energy sources, serving over a billion people in 50 countries and cutting annual CO2e of 1.25 billion tonnes (or a quarter of current emissions). The desert regions of northern Africa alone could satisfy 500 times the electricity demand of all EU countries.

The concept is for the construction of a €45 billion high-voltage direct current (DC) grid that could transfer the electricity produced by Saharan and North African solar installations to consumers thousands of kilometres away. The scale of the grid would overcome the intermittency inherent in wind and solar, by deriving power from wherever it was generated and delivering to wherever it is needed. For example, power generated through African solar power during the day could be stored in Icelandic hydropower stations, to be delivered, as needed, at night.

The potential of the Super Grid has led to increasing support for renewables. While many EU member states have been loathe to give domestic renewables priority access to their existing grids, they seem more comfortable with the idea of integrating renewables into a new Super Grid. Since 2002, Trans Mediterranean Renewable Energy Co-operation (TREC), has promoted the DESERTEC concept of building Concentrating Solar Power (CSP) plants in the Sahara desert to provide clean renewable electricity, in a proposal called the Mediterranean Solar Plan. TREC have successfully advanced the concept, combining CSP with wind farms and other renewables to transmit power to Europe from the Middle East and North Africa, via high voltage direct current (HVDC) cables across the Mediterranean. In 2008, the Mediterranean Solar Plan finally gained the backing of French President Nicolas Sarkozy and has also gained the backing of Germany and the UK. The key element is a new common framework for the sustainable development, by 2020, of 20 GW of new capacity in solar and other renewable energies in the countries around the Mediterranean Sea.

Dr Czisch of Germany's Kassel University has also proposed a Super Grid based on 70% wind power, supported by biomass and hydropower plants for baseline and storage capacity. The Czisch Super Grid would cover a region stretching from Siberia to North Africa, where the wind is most constant. There are about 1.1 billion inhabitants in the supply area, with a total annual electricity consumption of about 4000 TWh. According to Dr Czisch, if the power stations and transmission system were installed over time, through the replacement of existing plants as they age out of operation, the annual investment cost for the base case scenario development would be €52.1 billion for the wind power plants, €16.2 billion for biomass power plants, €6.4 billion for the HVDC transmission system and €2.7 billion for solar thermal power plants, totalling €77.5 billion. This is 0.6% of the EU's 2002 GDP.

High-voltage direct current (HVDC) transmission

Key to the implementation of a Super Grid would be the transformation of current transmission systems from a alternating current (AC) basis to a high voltage direct current (HVDC) network. While AC lines are the current international standard, they leak energy at a higher rate than DC. HVDC lines are three times as efficient, making them more cost effective over distances above 50 miles. Wind turbines and solar panels produce DC power, so using HVDC lines would eliminate the need to convert the power to AC at source. DC lines are also more effective in transmission underwater than AC lines, so it's a natural for offshore wind projects. These days HVDC-transmission lines are a more efficient way to move electricity over long distances, able to carry more power for the same thickness of cable compared with AC lines. As more renewables are brought online, this cost would go down.

Whether you're looking at an HVDC Super Grid in Europe to manage renewables across national boundaries, or one to link utilities within the US, such a grid would replace existing transmissions lines in an effective, low carbon-promoting, balanced power environment. Of course, the concept of an HVDC Super Grid is an international high-level attempt to address current power demand and supply issues but there are a number of other approaches which could be used in conjunction with such a development, or at a local, regional or national level.

The Smart Grid

One of the most effective ways that we could improve the reliability of our grid systems is through 'grid intelligence'. The majority of grids today lack the information, control and resource options to efficiently manage power losses from transmission (through transformer and line losses), as well as a range of power sources at different frequencies, voltage and reliability. The concept of the Smart Grid is to use the latest technologies in order to streamline day-to-day management of operations and to improve asset management processes (the range of generation types and locations that a future clean energy power market will need to use).

The US National Energy Technology Laboratory (NETL) defines the characteristics of a modern grid as:

- self-healing
- something that motivates and includes the consumer
- able to resist attack
- providing power quality for 21st century needs
- accommodating all generation and storage options
- enabling markets
- optimising assets and operating efficiently.

A well-planned Smart Grid would provide all of these.

On an operational level, using smart technologies would improve load factors, lower transmission losses and improve system reliability, as well as help to prevent power loss, black- and brownouts. Giving operators real-time information from advanced sensors would enable them to understand the state and condition of the grid at any given time, allowing them to plan around potential problems. Such information would enable the repair of equipment and generators before they fail and even help in managing the workforce. Not only would this improve reliability but could result in major savings in terms of the cost of blackouts to the economy.

As governments worldwide face increased pressure to produce renewable power generation on a mass scale, the grid will have to handle increasing amounts of electricity from intermittent and unpredictable sources, such as wind and solar. The proposed smart grid solution would use communications and automated control systems to smooth out erratic supplies by drawing on stored energy resources, balancing supply across a wide area, managing end-use demand and bringing networks of distributed generators online. In addition, advances in computer modelling will improve predictions of wind and sun availability, which will, in turn, improve forward planning.

The vision of an automated system that would replace existing grid management with a smart grid, is no less profound or widespread than that of the telephone, the personal computer or the internet. Smart grid management would rely on the latest digital technologies, requiring automated analysis of problems and automatic switching capabilities similar to internet technologies, such as the routers sold by Cisco Systems that break messages into packets and send them over different routes to relieve congestion, only to reassemble them when they reach their email destination.

Grid intelligence is increasingly being recognized as a real, long-term solution. It means that a mixture of the right hardware and software should be able to square the circle between fluctuating supply and fluctuating demand. For a long time generators have rewarded major customers for shifting energy demand to off-peak periods, whether that's been under contract or ongoing negotiations. Now, with the benefit of communications technologies, this could become the norm, rather than the exception and could be extended right down to the home user. The potential benefits will be substantial, not just in terms of increased reliability, but also in terms of lower costs. Reducing peak demand will reduce the need for investment in new generating capacity. More uniform voltage levels will reduce the number of customers receiving, and paying for, excessive power delivery.

McKinsey's report *How the IT Sector Can Save Money* states that use of sensors in grids would monitor the distribution of power more efficiently and help reduce losses. One grid in India that used information and communications technologies to monitor electricity flows, reduced its losses from the transmission and distribution of power by 15%. In India, where the generation of electricity accounts for almost 60% of emissions, reducing transmission losses by 30% could save the equivalent of €9 billion a year. Globally, a more efficient grid could generate €61 billion in energy savings and abate 2.03 metric gigatonnes of emissions by 2020.

Smart Grid technologies could also potentially deliver savings by making better use of available grid infrastructure capacity. Most grids are designed to meet the highest expected power demand, meaning that some parts will lie idle most of the time. In the US, for example, only 53% of power generation capacity is used, while just 50% of the transmission network and less than 30% of the distribution network is taken up. Under a Smart Grid scenario, digital technologies would reduce the need for 'just in case' infrastructure. If the grid were better balanced, demand could be met with far fewer plants. Studies have suggested that the deployment of smart grid technologies could eliminate over $100 billion in peaking infrastructure investments in the next 20 years.

Sensors in power lines give grid operators real time information about how much stress the power lines are under, thereby allowing higher levels of utilization. Furthermore, if digital technologies are infused throughout the grid it is possible to adjust end-use demands to suit available supplies, or bring local generation online to reduce the stress on power lines. Smart new energy devices and systems will need to evolve to relieve the pressure on overloaded grid infrastructure, reduce power costs, improve grid reliability and security, as well as accelerate the use of clean power generation. If these smart technologies are integrated throughout the economy, then smart equipment, appliances and buildings will have the capacity to power down, bring clean power options online or just help stabilize the demands of the grid.

A Pacific Northwest National Laboratory study calculated that smart energy technologies could lead to net present value savings of between $47–117 billion in avoided additional grid infrastructure costs in the US over the next 20 years. These savings do not take account of the capital costs of deploying new technologies but the report states that smart appliances costing $600 million would provide as much reserve capacity to the grid as power plants worth $6 billion.

Demand management is also going to play a critical role in the Smart Grid. Dynamic demand technologies are increasingly under consideration as a means of both cutting carbon emissions and balancing demand on the grid. Such technologies, integrated into a broad range of appliances, respond to system load pressure and can switch themselves on and off in order to reduce peak-load demand. For example, if a fridge is operating at a set temperature, but can maintain that temperature for six hours without further external power, it will stop drawing power if the system is overloaded.

There are a range of companies now working on technology to allow electrical equipment to communicate with smart devices and the grid in order to lower energy consumption. French company Watteco is developing devices for the home which warn consumers when their consumption is high, while Welsh company DeepStream Technologies is building circuits relying on sensor technology, which can be used in automating buildings, monitoring and controlling energy at the point of use within all electronic circuits.

Distributed generation

One option in both managing demand and integrating renewable power into the grid is to decentralize generation. The conventional wisdom says that in order to keep a power grid stable with regard to frequency and voltage, flows of power into and out of the grid must always be equal. If we deploy localized energy at the point of use, this no longer becomes an issue. It is possible that the combination of an ongoing upgrade of the grid combined with wide deployment of distributed generation, could result in higher power efficiency, lower costs, security of supply and a widely balanced network of generation.

The UK Office of the Gas and Electricity Markets (Ofgem) estimates transmission and distribution losses at between £600–700 million per annum for the UK alone. Under a highly distributed generation scenario, research carried out for the UK DTI identified possible savings of almost £35 million ($63.9 million) by 2020, arising from reduced network operation costs.

One of the criticisms of distributed power generation has been that while it can work very effectively for industry, on manufacturing sites, on farms, at leisure centres, it can be very difficult for individuals living in large cities to have any effective role. However, given technological advances it is now possible for utilities to operate a range of different small-scale power plants and combine them to provide power at point of use.

The next opportunity is to develop decentralized local energy generation networks, focused on industrial areas, housing groups or leisure centres. By setting up cogeneration plants near such large heat users, we could increase generation efficiency, cut tranmission losses and take demand off the grid.

Perhaps the most controversial opportunity is micro-generation, which has typically taken off in markets with a feed-in tariff. This simply means that utilities must buy all renewable electricity generated at a fixed price. For those with means of renewable energy generation at home and a meter, they generate their own power, draw down from the grid when needed and sell their surplus power to the grid as necessary.

Not everyone agrees that decentralization or micro-generation is entirely beneficial. Current power industry regulations largely derive from the unquestioned belief that centralized generation and distribution is optimal, without questioning whether or not the changing environment requires a different model. Some proponents of the more centralized model suggest that supplying energy is a public service that should be shielded from unpredictable market forces. There are concerns that decentralized generation could lead to large price fluctuations, or to potential supply disruptions caused by lack of oversight from a centralized grid.

If distributed generation is going to make a grid level contribution to power generation, a Smart Grid will be required. There's no question that digital management technologies that flexibly and automatically balance mass-scale intermittent generation with local distributed generation, energy storage and control of end-use demands, combined with effective forecasting and computer modelling could make such sources more predictable and therefore making them easier to integrate.

PART 2: BEYOND POWER GENERATION

There is far more to cutting carbon emissions than simply power generation. We also need to focus on minimising energy use in manufacturing, computing and buildings, as well as how we can evolve our transportation systems. Perhaps most importantly, we need to take a closer look at how we manage our current carbon stores – forests, for example.

One of the most effective tools we have, both to cut demand and lower emissions, is an increase in energy efficiency. Energy efficiency has often been seen as the ugly sister of renewable energy, but there is nothing ugly or unglamorous about saving money, reducing energy costs and lowering emissions. While the market tends to focus on investment in renewables as a means of cutting carbon, there is growing evidence that investment in 'negawatts', a term coined to describe a megawatt of power avoided or saved from use on the energy grid, will provide a better return.

At a consumer level, the modern economy clearly operates in ways that are wasteful and inefficient. Proposals to ban outright incandescent light bulbs and plastic bags, for example, are steps in the right direction. Highly visible, such actions serve well to underline the need for serious changes in lifestyle. There are still contradictory messages out in the market, however, and these need to be addressed.

At a policy level too, contradictory drivers exist. Increasing energy efficiency in buildings would be of clear benefit to society as a whole, but the disconnect between landlord/developer and tenant (where the developer pays for new technologies but the tenant gets lower bills and emissions) has slowed implementation. Clear frameworks and timetables for the phasing out of inefficient and carbon-intensive practices will help individuals and businesses better plan for the future.

Energy efficiency – the new fuel

The energy efficiency opportunity is immense – and this is something that the IEA calls the 'fifth fuel' after oil, coal, gas and nuclear. According to the McKinsey Global Institute's May 2007 report, *Curbing Global Energy Demand Growth: The Energy Productivity Opportunity*, increased energy efficiency is the biggest and most cost-effective lever to address GHG emissions.

The McKinsey report focuses on increased productivity (increasing the economic output per unit of energy input) but that's still talking about more bang for your buck, meaning lower power consumption for the same unit produced. According to the report, this could deliver up to half of the global GHG reductions required to cap levels at 450–550 parts per million. The report also highlights the fact that energy efficiency techniques are the only technologies with a material impact in terms of emissions reductions that actually save money.

McKinsey's research estimated that a global investment push into available energy-efficiency measures could halve the current projected growth in energy demand. By choosing more energy-efficient cars and appliances, improving insulation in buildings and selecting lower-energy-consuming lighting and production technologies, developing countries could cut their annual energy demand growth by more than half from 3.4% to 1.4% over the next 12 years. This would leave energy consumption some 22% lower than it would otherwise have been – that is, an abatement equivalent to the entire energy consumption of China in 2007.

Much of the anticipated energy demand growth globally could be met without building new power stations. By focusing on energy efficiency improvements, at a cost of $90 billion a year, most of that increased demand could be met by current capacity. The report concluded that investment in energy efficiency of about $170 billion a year to the year 2020, would yield a return of about 17%, or $29 billion. This would reduce global oil consumption by 21 million barrels a day, from the 2007 levels of 86 million barrels. Around half the total upfront investment required, $83 billion a year would need to be spent in industry, $40 billion in homes, $25 billion in transport and $22 billion in the commercial sector.

Tools to achieve these increases in productivity include the removal of fossil-fuel subsidies (it has been estimated that in the US, fossil fuels have received 60% of $700 billion in federal incentives between 1950–2006, oil at 46% and gas at 14%) as well as tighter energy efficiency standards for appliances, electrical equipment and new buildings. They also suggest focusing on higher corporate standards and investment in energy intermediaries, known as ESCOs or energy service companies.

While governments are exploring new policies for increasing energy efficiency, the McKinsey report makes a strong commercial case for immediate private sector investment in increasing energy productivity. Encouraging the private sector to see financial benefits in implementing climate change abatement measures can only prove a powerful engine for change.

Increasing industrial efficiency

Increasing concerns about carbon emissions are forcing corporations to rethink their approach to energy management. While many of the leaders of the energy-efficiency market drive are also at the forefront of the renewable energy market, investment in energy efficiency can deliver greater carbon reductions and financial return than investing in renewables.

Due to a difference in capital investment it could take 20 years to achieve payback on a wind turbine, whereas it would only take just over one year through energy efficiency. On a wider environmental point, businesses can reduce up to three times the amount of CO_2 for every £1 invested. This comparison shows that energy efficiency can provide a greater economic and environmental reward. What this means for an average business is an opportunity to invest relatively small amounts of capital in exchange for significant returns, a crucial requirement for company boards demanding to see tangible results.

The key to energy efficiency is measurement, based on the doctrine that what gets measured gets managed. Companies need to develop projects tackling energy efficiency in buildings and computer use, that can range from heating, cooling and lighting projects to complete reassessments of data centres and facilities management.

Businesses, from retailers to manufacturers, need a constant gauge of their energy usage to find out which parts of the business are the main offenders in terms of wastage. The critical advantage is having the know-how to determine customers' technical requirements and thereby identify and evaluate opportunities for energy efficiency measures. Then, using intelligent IT systems, the emissions profile of a building, or a network of buildings, can be addressed. It's the communication of information that is central to this process; information is what drives return on investment.

Energy efficiency is proving to be one of the most cost-effective abatement options to meet long-term emissions reductions, but beware a simple comparison of the costs of different options. Businesses need to realize that long-term changes in mitigating emissions and costs will also provide a positive impact on general operations and the economics make it clear that investment in negawatts rather than megawatts will provide a greater financial and environmental return.

Information technology

Information technology has a unique role to play within an energy efficient economy. By improving the efficiency of its own systems and power consumption, it can cut energy use dramatically within business. At the same time, it provides an opportunity to help transform the wider energy environment, through the implementation of demand/response technologies and the Smart Grid (see page 112). At the most simple level, this means enabling utilities to manage their customer information and power consumption through the installation of smart meters, or optimizing grid operations through the use of IT. Intelligent communications between a utility, its power transmission and distribution centres and end point devices requires a robust network. And the natural home of such a network would appear to be the existing network communications infrastructure.

There are two different kinds of energy use:

- in-use energy – power drawn by a specific device;
- embedded energy – energy used in manufacturing, distribution, packaging and disposal.

In IT, significant changes can be made by increasing the efficiency of in-use power and a range of technology companies – including Google, Dell, Sony and Yahoo – have all committed to energy savings through the use of increasingly energy-efficient technologies.

The price of electricity in the EU has risen an average of 31% since 2000, so the implementation of stand-by modes, power saving, efficient CPUs can make a significant difference to both costs and emissions. According to Hewlett Packard it saved 7.9 million kWh of electricity in 2005, purely by resetting 183,000 monitors worldwide to enable energy savings after 20 minutes of inactivity.

The industry is aware of the problem. In 2007, The Green Grid was launched – a consortium of IT firms, including Microsoft, Intel, Dell and Sun Microsystems, hoping to develop a set of standardized metrics to allow companies to measure and compare the energy efficiency of different servers, storage systems and networking equipment. It has proved difficult to develop global standards for measuring these due to the enormous range of players in the market.

While the consortium could be accused of dragging its feet, it has successfully launched a measure for assessing the efficiency of data-centre cooling and power supply units in the form of its Power Usage Effectiveness (PUE) metric; it is now working on standards for measuring the efficiency of the IT equipment itself and hopes to have draft versions of standards for completed efficiency metrics ready by the end of 2009.

A 2007 report by the research company Gartner, outlined the total amount of CO_2 emissions from the IT sector – roughly 2% of global carbon emissions, a level set to rise rapidly. Research from the Global eSustainability Initiative (GeSI) and the Climate Group found that, in 2007, electronic gear (meaning PCs, peripherals, telecoms and data centres), emitted 820 million tonnes of CO_2e – and this could potentially increase to 1.4 billion tonnes by 2020.

At the same time, IT could also provide the manufacturing sector with some easy wins. The use of information and communications technologies to optimize the energy efficiency of motors in China's plants, for example, could cut emissions by 200 million tonnes a year. The annual energy savings would amount to €8 billion, to say nothing of an additional €4 billion in the value of the emissions (at a carbon rate of €20 per metric ton). The value at stake globally in using these technologies to optimize the energy efficiency of such motors would be about €68 billion (€53.7 billion in energy savings and €14.7 billion in carbon savings). By 2020, this opportunity could reduce emissions by 0.68 metric gigatonnes annually.

The appropriate use of IT also has the potential to radically influence transport patterns (aside from the energy industry, the sector with the highest CO_2 emissions), as well as to influence our overall resource usage. On a basic level an increased use of video-conferencing and/or remote working would impact on travel patterns.

According to the McKinsey report How IT Can Cut Carbon Emissions the demateralization of physical goods and services – such as using the internet through video conferencing, online shopping, and so on – would cut emissions significantly, by 0.5 metric gigatons a year, but far less than the 7.3 metric gigatons in annual emission reductions from improved energy efficiency in factories, buildings, electricity grids and truck fleets.

Another area where IT could further transform the wider information technology market itself is through the roll-out of cloud computing, or the use of virtual servers. This is where applications and software are provided as a service over the internet, based on one set of highly optimized servers. The remote support of applications and software through the usage of virtual servers could have a significant impact on the overall usage of servers and their concomitent power consumption. This could be accompanied by a decrease in data-centre energy consumption of up to 70%, because more effective centralized power management could be implemented.

While the emissions from IT industry energy consumption do not compare to the power or transportation sector, ineffiency affects every industry within the developed economy. At the same time, as industry generally looks to reposition itself in a low carbon, energy-efficient economy, the IT sector has an opportunity to help focus and develop new approaches.

Of course, one of the most effective and immediate ways in which IT can be deployed to cut carbon emissions is in its use in buildings power management.

Buildings

According to the 2007 UNEP report *Buildings and Climate Change – Status, Challenges and Opportunities*, more than one-fifth of present energy consumption and up to 45 million tonnes of CO_2 per year could be saved in Europe by 2010, by applying more ambitious standards to buildings. Achim Steiner, UN under-secretary general and UNEP executive director, stated in the report that by conservative estimates, the building sector world-wide could deliver emission reductions of 1.8 billion tonnes of CO_2. A more aggressive energy efficiency policy might deliver over 2 billion tonnes or close to 3 times the amount scheduled to be cut under the Kyoto Protocol.

'More than one-fifth of present energy consumption per year could be saved in Europe by 2010, by applying more ambitious standards to buildings.'

According to some research, emissions from buildings (and their use of electricity and energy) are responsible for 40% of all global emissions and up to 80% of total GHG emissions in our cities and towns, However, the sector has some of the greatest potential for transformation to a low carbon economy.

There are a number of critical issues to address when approaching the problem of cutting emissions in the built environment, including:

- the disconnect between developers and tenants in terms of implementing energy saving technologies;
- a fundamental failure to integrate developments into a wider social plan;
- as most building stock we'll use in the next 50 years already

exists, there is a need to implement changes in existing building stock.

Significant technical opportunities exist to cut emissions in buildings, especially when in development. One of the key carbon/energy solutions for buildings, especially in the commercial sector, lies in something as simple as lighting and temperature controls. Advanced integrated wireless lighting control systems, for example, mean that lights switch off when a room is empty; similarly, temperature controls can respond to real-time weather conditions, positioning blinds where they're needed to provide optimum light and shade – these help ensure energy is only used when and where needed and won't compromise the comfort of the building's inhabitants.

One of the key issues to be overcome in the building sector is how to satisfy the interests of both the developer and end-user of the building. In the life-time of an average building most energy is consumed, not for construction, but during the period when the building is in use. That is, when energy is being used for heating, cooling, lighting, cooking, ventilation and so on. Typically more than 80% of the total building energy consumption takes place during their use and less than 20% during construction.

Green or low carbon development has made buildings a flagship issue, but the reality of developing new green building stock or retrofitting existing stock is quite difficult. In new builds, building controls and ways of increasing energy efficiency should be a key component of development. Taking advantage of natural daylight would decrease the need for artificial light, for example. The orientation of the building itself could mean having main living spaces facing south, to take advantage of sun and shade, or, where buildings need cooling, avoiding west facing windows.

Heating systems, especially in relation to leisure centres, hospitals or industrial estates, could be combined with power systems to increase efficiencies and passive solar heating could take pressure off heating demand. Insulation is one of the key ways of keeping costs low, as is lighting control. The enormous number of windows included in many modern designs can also be an issue – double glazing can cut heat loss, or developers could look at building-integrated photovoltaics (BIPV) or even solar reflectors in high-temperature locations.

According to McKinsey, in the US alone using such tools as building management systems (BMS) – managing lights, heat, ventilation and air conditioning (HVAC) systems, fire safety and security equipment and even communication networks – could cut the commercial-building sector's energy bill and associated emissions, by nearly 30% a year. Globally, smart buildings could cut emissions by 1.68 metric gigatons a year.

It would make a significant difference if developers would install energy-efficient boilers and other forms of renewables. Active solar heating, wood, waste or biomass fired plants, even ground source heat pumps, could be considered for larger installations or blocks of flats. Air-conditioning systems can also be optimized, using natural ventilation and even internal greenhouses to manage heating and cooling. Further suggestions include using materials that cut down on condensation, ensuring the provision of adequate daylight, sound insulation and even the provision of private or shared outdoor space. These can not only be used to improve heat management but can contribute to a more pleasant environment. As ever though, we hit the problem of who is paying for the installation of such features (the developer) against who benefits from the savings (the tenant or purchaser).

One area that shouldn't be ignored is the importance of water recycling. Certainly in the UK, water is used for drinking, washing, watering the garden, industrial processes and more, and all UK water is cleaned to a drinkable level, using unnecessary amounts of energy. The average UK person uses 50 litres of water a day in flushing the lavatory, so why not recycle water used in washing for flushing? There are now complete bathroom control systems for the non-domestic market that regulate water supply, lighting and ventilation, supplying services on demand. Water butts on roofs could also enable large developments to source excess water to use outside.

Finding a way to handle CO_2 emissions associated with buildings is critical. The largest building markets are found in the developing world, in China, India and Brazil. China is building around 2 billion square metres a year – the same as a third of Japan's existing built area. While there is an opportunity here to leapfrog existing design and build methods (much as telecommunications leapt straight to mobile networks in many developing markets) that process must be supported and that means that we need to address behavioural, organizational and financial barriers to change.

We have an opportunity to take a broader look at wider planning issues. Where possible we should look at landscape strategies to reduce the urban heat island effect. According to the US Environmental Protection Agency, the annual mean air temperature of a city with 1 million people or more can be 1–3°C (1.8–5.4°F) warmer than its surroundings. On a clear, calm night, however, the temperature difference can be as much as 12°C (22°F). This can lead to increased energy consumption for heating and cooling, increased carbon emission and impaired water quality. Strategies for managing this can range from the creation of gardens on roofs, or reflective roofs and increasing tree cover and vegetation.

According to the WWF using existing technologies to help plan sustainable communities during city planning could save 22 million tonnes of CO_2 in the EU-25 alone.

Heating and cooling still have the largest impact and a change in baseline temperature of 1°C (33.8°F) can provide savings of around 7% on the energy used to generate that heat. It doesn't always work, however. Added insulation is seen as a prime means of changing energy drain in buildings, yet in many existing buildings it can be difficult to add further insulation. In the UK, as Green Alliance pointed out in a 2007 report, 43% of existing

UK housing stock have solid walls, which makes cavity wall insulation impossible.

The implementation of standards for new development does little to address the problem of existing building stock. For example, UK legislation for zero-carbon buildings focuses initially on new build – homes by 2016 and non-domestic buildings by 2019 – which, at best, accounts for 1% of building stock on an annual basis. The UK has already built more than 70% per cent of the building stock that will be available in 2050, so reducing the carbon footprint of the existing built environment needs to be the highest priority.

Given that only a certain amount of refitting of existing stock is possible, this brings us to the issue of behavioural change. That is, changing how we use our buildings and what we expect them to provide. Most of the energy used in buildings is to maintain 'comfort' through heating and air conditioning. If we are to succeed in changing our expectations, this could have a significant impact on energy use.

The July 2008 issue of *Building Research & Information* explored the different ways in which we can address the issue of comfort in buildings, especially in relation to climate change. Editor Richard Lorch noted that the most significant climate change has been in the indoor climate – the range of indoor temperature has decreased significantly over the past 40 years.

According to author Elizabeth Shove, the standards, regulations and (efficient) technologies for indoor climate control have often have the opposite effect of their intention: they increase energy usage. For example, in the UK consumers are being encouraged to keep the heating down to 21°C (69.8°F), but two generations ago that would have been unheard of comfort.

The current climate crisis demands a concerted effort to understand how our definitions of appropriate comfort have both a history and a future, enabling a shift from automation and predictability towards adaptability and resilience. In the current energy framework, individual expectations are met as asked, instead of engaged with. We need to rethink how we expect to live. It is common for people to now live in areas of the world where the temperatures are too hot or too cold to make life comfortable. It has become the norm to change our environment to suit ourselves, which may not prove the best approach to addressing the problem in the longer term.

Transportation

Transportation accounts for around one-fifth of global GHG emissions, although that figure can run as high as 30% in developed economies. More importantly, emissions from transport are growing faster than any other sector. The number of new vehicles produced each year has risen from 5 million in the 1940s to more than 55 million in 2007.

Cutting emissions in the transport sector can be less cost effective than in other sectors, as there are so many different components to model. There is high price sensitivity among car and truck drivers, growing demand for personal and industrial transportation, as well as in introducing new low carbon fuels which can often be more expensive, less efficient or be difficult to distribute. At the same time, as easy to access oil reserves diminish, the quality of the oil can mean additional refining (and therefore energy consumption) is required to transform it into useful fuel. All these aspects make transportation one of the most difficult sectors to address.

IT can help in logistical terms, as technology could potentially cut annual global emissions from trucks by around 1.52 metric gigatonnes a year (according to McKinsey research). In Europe, companies with six or fewer trucks own 60% of the fleet. If smart transportation technology were to manage the flow of truck traffic in Europe – as it already does on some German highways – truckers could work together to optimize their loads. While the management of traffic flow will have an impact, the key issue for the time being remains how best to manage and mitigate the emissions from the road vehicle, shipping and aviation sectors and typically, that means managing fuel.

Road vehicles

A 2008 report from Norway's Center for International Climate and Environment Research (CICERO) suggested that not only was road traffic by far the largest transport sector contributor to emissions, but could make up to 75% of all warming caused by transport over the next hundred years, if left to grow unchecked. There are concerns about both aviation and shipping, but their relative contribution is far smaller.

While there are opportunities to cut the growth of fuel emissions, through increases in vehicle efficiency, lower carbon fuels and in behavioural change through travel reduction, there are few options available today which can make a significant difference. While the average road vehicle has seen great technological leaps in terms of materials use, lower resistance tyres, direct fuel injection and more, this has yet to result in major decreases in fuel consumption. That, in turn, means little significant difference in emissions levels.

Petrol cars are more efficient today than they were a decade ago, but increases in size and weight have tended to outweigh this. Improvements that have been seen in CO_2 emissions have tended to be related to an increase in the use of diesel which, although there are some concerns about particulate emissions from the engine, are more efficient than their petrol-fuelled brethren.

Many believe that fuel cells and the hydrogen economy lie at the heart of any post-fossil-fuel energy architecture. Although they have been around for 150 years and their performance is not in doubt, high manufacturing costs and low reliability have meant that they have, as yet, failed to capture a commercial market. There are some markets where fuel cells are in customer trials, such as for auxiliary power units (APUs), uninterruptible power supplies (UPS), backup power and more. However, in the vehicle market there continue to be problems with fuel storage, weight, vehicle range, reliability and durability so, for the moment, they're not going to help.

In the long run, fuel cells may be the best option given energy supply, security and environmental problems, but we need to look at transformational changes that we can make today. That means that we need to focus on increasing efficiency, alternative fuels and electric vehicles.

Increasing fuel efficiency

The EU was amongst the first international communities to propose direct regulation of CO_2 emissions, rather than fuel consumption. In order to avoid mandatory cuts, the European car industry agreed a voluntary arrangement which was to see new cars reaching an emissions target of 140 grammes of CO_2 emitted per kilometre (g/km) by 2008. By and large the car industry failed to achieve this, with the average emissions from new cars today sitting at around 160 g/km.

The EU has now included regulation of vehicles emissions in legislation, and the target is for all manufacturers to reduce vehicle emissions to 130 g/km by 2012. This is expected to result in a 19% overall reduction in car-related CO_2 emissions. The car lobby continues to argue that such mandatory targets create financial problems for an industry already in trouble, but the original voluntary agreement did have some effect and there are already cars available that beat the new EU target. Within the EU, BMW, Citroen, Daihatsu, Fiat, Ford, Honda, Peugeot, Renault, Smart and Toyota all have produced models which emit less than 120g CO_2/km.

There are a range of ways to cut emissions, through new engine management systems, powertrain improvements, the introduction of turbochargers and through the introduction of hybrid electrical systems. An EU project is even developing a new stop-start system to replace alternators, which will result in no fuel consumption, gas emissions, vibrations or noise when standing still. If the industry as a whole is to meet the 130 g/km target, there will be a direct impact on those manufacturers who develop the larger, luxury, less efficient vehicles. Cost remains an issue, however – estimates have been made that the cost of compliance in a mature market, like the EU, could add between €1–2,000 to the average price of a car.

In the US, the focus remains on fuel consumption, rather than emissions levels. The car industry received a federal bailout of $17.4 billion in December 2008 and agreed to clean up the industry in return. Congress has passed legislation to increase average fuel consumption to 35 miles per gallon (mpg), from existing fuel economy standards of 27.5mpg for cars and 24.1mpg for light trucks by 2011. US President Barack Obama may have promised to make the White House fleet plug-in hybrids, but that is a long way from implementing strong standards on US cars. The US economy is in the same trouble as everyone else, the average US citizen is used to cheap fuel and the prices continue to go up.

In developing economies there is an opportunity to cut straight to efficient vehicles through the implementation of hybrid or electric vehicle technology at a relatively early stage of market development. China, for example, has annual growth rates of 10% in the numbers of cars on the road and it has been estimated that this could mean 50 million new cars on the road every year by 2050. This provides China with a unique opportunity to look at alternatives in sustainable transport solutions.

The Chinese are keen to slow oil consumption, which is already at 40% of its primary energy sources, and, as far back as 2004, set a national fuel consumption standard. China is exploring the option of flexible fuel vehicles, compressed natural gas-fuelled vehicles and grid independent electric vehicles. It is also looking at fuel cell, grid connected electric vehicles and pure electric vehicles.

Biofuels

There are two key types of biofuel:
- bioethanol: produced by fermenting and then distilling starch and sugar crops such as maize, potatoes, wheat, sugar cane and even fruit and vegetable waste
- biodiesel: made from plants that contain high amounts of vegetable oil, such as palm oil, rapeseed, soybean and jatropha

Biodegradable outputs from industry, agriculture, forestry and households can also be used for biofuel production e.g. timber, manure, sewage and food waste. The use of biomass fuels can, therefore, also be seen to contribute to waste management as well as fuel security and climate change mitigation.

The last few years have seen a massive increase in the amount of ethanol and biodiesel used in vehicles around the globe. While Brazil has had a significant biofuel mandate for a couple of decades, it's only in the last few years that we have seen such developments in Europe and the US. Biofuels derived from plants gained in popularity as concern grew regarding carbon emissions, as they are considered carbon-neutral over their lifetime.

Growing concerns about the sustainability of biofuels have increased however, especially as so many food crops have been used in its production. Changing weather conditions are affecting crop yields around the world, while at the same time a growing demand for biofuel is adding additional upward pressure to food crop prices.

'Agflation', a term coined by Merrill Lynch to describe the continuing upward pressure on food costs, has been driven by a combination of bad weather affecting food production, increasing quantities of food consumption globally and the growing focus on using biofuel as a replacement for fossil fuel in many markets. Cargill, the US-based conglomerate, has invested $1 billion in ethanol and biodiesel production, but it also supplies a quarter of the meat consumed in the US, where margins have been hit by higher feed costs. In June 2008, new chief executive, Gregory Page, warned that biofuel mandates and other incentives could distort the allocation of land, with the potential to create food shortages around the world in the wake of 'weather-related crop problems'.

The US and Brazil are the largest producers of bioethanol. In the US, corn has risen dramatically in cost as ethanol production increases, while Brazil's rainforest is being affected, as developers raise their production of sugarcane. In 2008, President Luiz Inácio

Lula da Silva was even forced to defend Brazil against charges that biofuel expansion was threatening food production and causing increased deforestation of the Amazon.

The production of biodiesel from palm oil is also of concern, as old forest has been cut down to produce increasing amounts of palm oil, all over South-East Asia. While technological breakthroughs may yet turn switchgrass, wood chips and other forms of organic matter into profitable and environmentally sound alternatives, they are not economically viable. There is a possibility that eventually demand from the food and biofuels industries could result in food crops beginning to track the price of oil.

All of this means growing dissension about the degree to which biofuel should be used in the transport sector. Reports from the OECD in November 2007 and the UK's Royal Society in January 2008 expressed concerns about the contribution biofuels can make to the climate change mitigation debate, as well as the potential for harmful environmental and social impacts.

Despite the inclusion of a 10% mandate for biofuels in the EU Climate Action and Renewables Package, there are signs of back tracking. The European Commission have attempted to bypass the issues surrounding use of biofuels by insisting that although they are seen as a key part of renewable energy policy and a crucial solution to growing emissions in the transport sector, biofuels should not be promoted unless they can be produced sustainably.

The main criteria in the proposals regarding biofuels is that no old forest, biodiverse grassland, wetlands or peatstock (carbon-heavy environments) should be used as raw materials in the generation of biofuels and that the overall CO_2 savings from biofuel production must be at least 35%. To calculate the emissions cuts, the entire lifecycle and production chain of the biofuel in question must be analysed.

This means that several biofuels will no longer meet the criteria, including biodiesel derived from rapeseed and bioethanol from corn, potatoes and rye. Palm-oil based biodiesel will meet the 35%

target, although producers will now have to prove that the feedstock was grown on land that was not previously forested. Most second-generation biofuels are expected to meet the 35% emission reduction target fairly easily, as the technologies involved in production are far more efficient than current methods.

In July 2008, the UK's *Gallagher Review* concluded that there is a future for biofuels, but it called for the introduction of biofuels to be slowed until effective controls are in place to prevent land-use change and higher food prices. The Review proposed that there should be an EU obligation to cut emissions in the transport sector through the use of a set percentage of biofuel used in fuels. The target begins in 2015 and rises to 1–2% of energy demand by 2020. To qualify to meet the target, UK biofuels will need to deliver high GHG savings from the use of appropriate wastes, residues, crops grown on marginal land, or feedstock, such as algae, that do not require agricultural land.

Cost is, of course, another key factor at play. Outside of Brazilian ethanol (because it has been part of the Brazilian fuel mix for a long time), biofuels cost more than other forms of renewable energy and without a separate minimum target for biofuels, they will not be developed. According to the Commission, this matters because GHG trends are worst in transport and biofuels are one of the few measures – alongside increasing fuel efficiency – realistically capable of making significant inroads on GHG emissions from transport. Of course, biofuels also take pressure off the oil price – a March 2008 analysis from Merrill Lynch reported that biofuels keep world oil prices 15% lower than they otherwise would be. In Brazil, gasoline prices didn't go up in two years because of competition from ethanol.

There is still a great deal of interest in second-generation biofuels, which use non-food crops or waste products to generate fuel. They usually involve a process where enzymes are used to break down the cellulose and lignin in plant fibres and turn that into sugar, but the process is expensive. Although there are pilot plants in development, most experts believe that we are 3 to 5 years from commercial operations. Another thermal method known as the Fischer-Tropsch process, which was originally used in the transformation of coal to a liquid fuel, involves heating waste to produce a synthetic gas or syngas, which can then be reconstituted to create a liquid fuel.

There has also been a great deal of interest in biodiesel, especially from sources such as Jatropha Curcus, a biofuel feedstock crop that can grow on wasteland. It provides an opportunity to reclaim such land and means that arable land can be retained for food. There is still some time to go before the first projects prove their viability, however, and there are growing concerns that in the developing world, Jatropha is nonetheless being grown on arable land, as it's possible that good cropland could be designated wasteland if the right deal is agreed.

Another possibility is the generation of biofuels from algae. Certain types produce substances that can be turned into fuels, using CO_2 as a feedstock. Algae fuel has attracted attention because it does not require farmland or forest but, at present, no one is producing it commercially and it is a long way from becoming a major fuel source. Estimates suggest that algae could produce almost 100,000 litres of biodiesel a year per hectare of land equivalent, compared to 6,000 litres a hectare for oil palm. Origin Oil is one company looking at several strains of algae, which can be harnessed in this way and in the long term plans to ship standardized containers of algae which customers can use to set up their own biofuel refineries.

But is the use of biofuels sustainable, not simply in terms of environmental but also in terms of social and human rights issues? Potential problems range from concern regarding its CO_2 and energy balance during production, to ecosystem loss, water scarcity, population displacement and increases in food prices. The use of ethanol has, in particular, come under attack following release of research that shows that it takes more energy to make ethanol than the end result contains. While other studies have shown the balance to be positive, the debate continues to rage.

Electric vehicles

The mass production of electric vehicles is still some way off but the concept seems both sensible and efficient. Electric motors are efficient energy converters, promising around 80% efficiency rather than the roughly 30% of petrol/diesel engines. Battery-electric vehicles (BEVs) can reduce car emissions to about 50% of current models. But the issue remains the selected combination of technology, cost and infrastructure. If the electricity comes from fossil fuel, it somewhat defeats the object.

Nissan plans to launch a battery-powered car in America in 2010, and by 2012 the Renault–Nissan alliance plans a complete range of electric vehicles in every large car-market. These new battery-powered cars, it claims, will work out less expensive than equivalent petrol models. For other original equipment manufacturers (OEMs), hybrids may provide a bridge to a pure-electric future. Chrysler has announced new hybrids and one BEV sports car and Toyota has said it will have hybrid versions of all its cars by the 2020s.

It's not all bad news for lovers of the luxury car. Pay-Pal co-founder Elon Musk's Tesla Motors is aiming an electric car at the luxury market. Tesla's current model, which is a two-door roadster, costs consumers in excess of $100,000 and the company is only just managing to match demand.

Model S is next on Tesla's zero-emission list, a five-passenger luxury sedan powered by a lithium-ion battery pack. It is expected to have a base price of about $60,000 and get about 240 miles per charge. The first sedans are expected to roll off the assembly line in late 2010. Future models are said to include a four-door sedan that will be unveiled early in 2009, followed by an economy car that will cost consumers 'below $30,000.' All of this hinges on improvements in Tesla's production techniques and other advancements in technology; Musk hopes to achieve production in the hundreds of thousands per year – basically enough for them to make it to most dealerships and more importantly to get potential customers off long waiting lists.

One of the things that has delayed the wide-spread role out of electric cars has been lack of infrastructure. Most users would need on-street or car-park recharging points, which don't exist in any number. They're expensive to build and the vehicles will need to charge for at least an hour. The concern is that a large number of electricity sub-stations will be required to recharge multiple vehicles from a single supply source and fast-charging systems are unlikely to be feasible in many public sites, let alone compatible with domestic power circuits. And, of course, there is the concern about the impact on the grid if charging suddenly peaked.

But things are beginning to change.

In 2008, French car-maker Renault and French utility EDF announced a government backed scheme to build a national electric car recharging network by 2011. The project will receive €400 million of state backing over the next four years, which has been personally guaranteed by President Nicolas Sarkozy. The two companies have agreed to develop a commercial project that will see the establishment of charging stations and the development of an electric version of Renault's Kangoo van and a new electric sedan called the Fluence.

There's also an alternative to a complete BEV infrastructure, developed by California-based Better Place. In the past, the focus has been on creating batteries that held a charge for distances of more than 400 miles (644 km). This is an expensive, challenging and more futuristic undertaking. Better Place plans to deploy lithium-ion based batteries, which are the most advanced in the industry, being the safest, smallest, lightest and most efficient available. By separating the battery from the car and making it a component of the infrastructure, the whole approach changes. The battery can be recharged and replaced simply through automated battery exchange stations. This approach allows Better Place to create an ecosystem that uses the battery technologies that are already available, making electric vehicles an immediately viable alternative to fuel-powered cars.

The first network the company is building is in Israel, where 90% of car owners drive less than 70 km (about 43 miles) per day and all major urban centres are less than 150 km (about 93 miles) apart, making electric cars an ideal means of transportation. The group also has an agreement with Denmark's DONG Energy. DONG Energy and Better Place are working on a project to give Danish consumers access to buying electric vehicles at low prices. Within the next few years, Better Place Denmark is to introduce environmentally friendly BEVs to the country. Better Place has also announced agreements with AGL Energy and financial advisor Macquarie Capital Group in Australia, to raise AUD $1 billion and begin deploying an electric vehicle network in Australia. With the world's seventh highest per capita rate of car ownership, the country has nearly 15 million cars on the road after adding over a million new cars last year. Other projects are under consideration in Japan, Canada, Hawaii and the San Francisco Bay area.

In August 2008, Spanish Industry Minister Miguel Sebastian stated plans to put 1 million electric or hybrid cars on Spain's roads by 2014 and he announced three working groups to develop infrastructure, boost sales and cooperate with Nissan in order to build electric cars. The plan will cost some €245 million but Spain is hoping to save between 5.8 and 6.4 million tonnes of oil over the three-year period that the plan is implemented.

The overnight charging structures in such a network would also provide a potential solution to Spain's wind energy storage problem. Spain has over 15GW of wind power installed and without sufficient demand, that power needs to be bled off the grid. In November 2008 high winds meant that Gridco Red Electrica de Espana (REE) had to pull 37% of its wind generation capacity off the grid in order to keep the network stable. It's possible that the widespread implementation of electric vehicles could contribute significantly to the amount of renewable power that a national grid could carry.

There is also an increasing focus on electric vehicles in emerging economies. India is home to the world's best selling BEV, the Reva (or G-Wiz) from Bangalore, while China's electric vehicle industry is enormous, with 9 million electric bicycles manufactured in 2005. Warren Buffet, one of the world's most respected businessmen, has invested $230 million in Chinese battery manufacturer and car maker BYD, which launched a mass produced hybrid electric vehicle at the end of 2008.

In Tianjin, an electric vehicle factory is under construction that will boast a capacity of 20,000 units per annum. When completed, the Tianjin–Qingyuan Electric Vehicle Company will be the largest electric vehicle manufacturer in the world and – it is worth repeating this point – it will be a Chinese company using Chinese technology, with plans to export half of its annual production to the US and Europe.

Whether pure-electric cars gain critical mass fast enough to halt transport's rising CO_2 emissions will depend on decisive, concerted efforts of OEMs, utilities and governments around the world. If the technology is to have a significant impact, it will have to show stronger growth than current hybrid technology – Toyota's Prius took nearly a decade from launch to sell 1 million cars.

Aviation

Of course, transportation is about far more than cars and trucks. According to the IPCC, CO_2 emissions from aviation – directly related to the amount of fuel consumed – have increased by 87% since 1990 and now account for around 3.5% of total anthropogenic contribution to climate change. It's estimated that this share will grow to 5% by 2050.

At the same time, the IPCC says the total impact of aviation is about 2 to 4 times higher than its carbon emissions alone, due to aircrafts' emissions of nitrogen oxides (NOx) and water vapour in their condensation trails, as well as the impact of where their emissions hit the atmosphere.

While aviation was not covered under the Kyoto Protocol, the European Commission has formally approved a scheme for including aviation in the EU ETS from 2012. Under the new proposals, airlines will be obliged to buy 15% of their required carbon allowances at auction, while emissions caps will initially be set at 3% below 2004–06 levels, rising to 5% for the period from 2013.

All flights in and out of the EU would be included, which should prevent any distortion of competition between EU and non-EU airlines. However, while many airlines seem to accept the importance of focusing on the necessary greening of aviation in a carbon-constrained world, no consensus has been reached on what changes to make or how.

One of the fundamental problems of the commercial airline industry is the difficulty in switching to alternative fuel sources. But given the 87% cumulative growth in CO_2 since 1990 against the EU's overall GHG reduction targets, it's clear that something must change. Although the air transport industry has made improvements to aircraft technology, the resulting reductions in GHG emissions have not been sufficient to compensate for the rapid growth of global air traffic, up roughly 50% over the last decade.

February 2008 saw the launch of a seven-year, €1.6 billion EU public–private partnership, aimed at helping the aviation industry to develop environmentally friendly technologies. The Clean Sky Joint-Technology Initiative is intended to support the technological developments necessary to cut aircraft noise in half and for emissions of CO_2 and NOx to be cut by 50% and 80% respectively by 2020.

The International Air Transport Association (IATA) said that 12m tonnes of CO_2 could be saved annually by improving infrastructure and operational inefficiencies. Director General and CEO Giovanni Bisignani said, 'With fuel making up 28% of operating costs, airlines have a $132 billion economic incentive to reduce fuel burn and CO_2 emissions'.

Some commentators doubt whether inclusion in the ETS will lead to a reduction in emissions in the air transport sector, especially given the growing demand for air travel. It is possible that the move will lead to compensatory reductions in CO_2 emissions elsewhere in the economy but it's important to note that the pressure to make changes within the aviation industry is not simply coming from policy makers.

In 2008, Standard Life Investments announced it would no longer be investing in airline stocks through its ethical funds, as a survey reported that 30% of its investors objected. This move highlights the debate on the role of investors in carbon intensive sectors and underscores the need for companies to be seen to be actively mitigating their contribution to climate change, as discussed in Chapter 3.

It is becoming increasingly clear that airlines must move towards controlling their GHG emissions profiles, whether the main driver of change is fear of the cost of emissions trading, saving money through increased efficiencies, or even shareholder appeasement. It might be preferable for an international system rather than solely an EU framework, but the first steps have to be taken at some point. However, there are problems to overcome. In 2008, Giovanni Bisignani warned that the EU's unilateral approach to adding aviation to an ETS is illegal and an ineffective way to deal with aviation emissions.

In 2008, he outlined a number of problems with the EU's unilateral approach, not least of which is additional costs of around $3.5 billion on an industry which is under drastic economic pressure and already operates in a highly fuel efficient fashion. Bisignani warned that the inclusion of aviation in the EU ETS will introduce commercial distortions that could hurt Europe's carriers, meaning it could be cheaper to fly to Singapore via a Middle East hub than flying direct. He called the ETS concept 'morally bankrupt' when the European Parliament suggested only that profits from the scheme 'should' be spent on environmental projects.

However, his key message is that the inclusion of aviation in the EU ETS is illegal. Over 130 countries have already stated their opposition to Europe's action. He said, 'Instead of cleaning up the environment, Europe is creating an international legal mess with a responsible airline industry caught in the middle. If Europe genuinely wants to take leadership on environment it must support the Group on International Aviation and Climate Change (GIACC) at ICAO, the only hope for a global solution. Europe is represented by France, Germany and Switzerland. They must be strong voices for effective global solutions'.

Shipping

Shipping has also come to the attention of climate change campaigners. A UN report from the International Maritime Organization (IMO) said that pollution from shipping was actually three times higher than previously believed. It calculated that annual emissions from the world's merchant fleet have already reached 1.12 billion tonnes of CO_2 or nearly 3.5% of all global emissions of CO_2. It also said that growing international seaborne trade and related fuel consumption would be likely to raise CO_2 emissions from ships by 30% to 1.48 billion tonnes by 2020.

By comparison, the aviation industry, which has been at the sharp end of climate change pressure, is responsible for about 650 million tonnes of CO_2 emissions a year, just over half that from shipping. Governments and the EU have consistently played down the climate impact of shipping, saying it is less than 2% of global emissions and failing to include shipping emissions in their national estimates for CO_2 emissions. Pressure is now expected to increase on ship-owners to switch to more efficient fuels and on the EU to include shipping in its emission trading scheme at some point in the future.

Shipping is responsible for transporting 90% of world trade, a level that has doubled in 25 years and is expected to grow by as much as 75% by 2030. Previous attempts by the industry to calculate levels of carbon emissions were largely based on the quantity of low-grade fuel bought by ship-owners. The latest UN figures are considered more accurate because they are based on the known engine size of the world's ships, as well as the time they spend at sea and the amount of low-grade fuel sold. The UN report also reveals that other pollutants from shipping are rising even faster than CO_2 emissions. Sulphur and soot emissions, which give rise to lung cancers, acid rain and respiratory problems, are expected to rise more than 30% by 2020.

While we need more work on analysing the impact of shipping on carbon emissions and how it might be managed, the industry will need increased efficiency, alternative fuels, or other ways of cutting emissions. At any rate, it seems certain that the sector will become part of the debate in future climate change negotiations.

Forests: critical carbon stores and more

There are three critical carbon stores on Earth – the atmosphere, the oceans and forests. While the attempt to cut emissions under an international treaty addresses the atmosphere, the oceans and forestry have been rather taken for granted.

The oceans have been absorbing CO_2 for millenia, but there are concerns that the increasing uptake is causing the oceans to become more acidic, slowing the potential uptake of CO_2. There are significant changes in the natural carbon cycle of the oceans and this can make it hard to track exactly how the oceans are being affected by anthropogenic climate change. There is considerable work to be done in understanding the role of the oceans in the carbon cycle. The role of forests however, is far better understood.

Simply put, living forests, wetlands and peatlands can sequester carbon emissions, while dying ones release previously stored carbon. With forests able to act as both carbon sinks and carbon sources, if we're ever going to have a chance to solve the problem of climate change, we're going to have to handle the impact of land-use change, land degradation and deforestion. So we need to take a closer look at forestry, and the opportunity of having naturally grown trees (standing forest as opposed to tree plantations) and soils to store CO_2.

The vegetation and soils of the world's forests contain about 125% of the carbon currently found in the atmosphere. When forests are burned, degraded, or cleared, the opposite effect occurs: large amounts of carbon are released into the atmosphere as CO_2 along with other GHGs (nitrous oxide, methane and other nitrogen oxides). The burning and conversion of the roughly 14 million hectares of forests lost per year (according to the Finnish Forest Industries Federation) releases more than 2 billion metric tonnes of CO_2 into the atmosphere, or about 20% of anthropogenic emissions of CO_2 and 25% of total GHG emissions.

Forests also provide ecosystem services (as discussed in Chapter 2), which are vital to the functional operation of the biosphere. Loss of forests through land-use change is the greatest driver of global biodiversity loss and a major contributor to global GHG emissions. The second report of the IPCC stated that since the Industrial Revolution, 136 gigatonnes of carbon have come from land-use change, while 270 gigatonnes of carbon have come from fossil-fuel and cement production. According to the World Resources Institute (WRI), land-use change accounted for 25% of all global emissions in the 1990s alone.

In the last 300 years, global forest cover has shrunk by approximately 40%. Forests have completely disappeared in 25 countries and another 29 countries have lost more than 90% of their forest cover. Every year, forest is lost for a range of different reasons: conversion for agriculture, wood fuel, ranching, palm oil plantation, population expansion, infrastructure growth and also from logging (both legal and illegal). This conversion of forest land to other uses is responsible annually for the release of up to 20% of all GHG emissions. However, as long as trees are perceived to be more valuable cut down than they are standing, deforestation will continue.

Ecosystem services

Stopping or slowing the rate at which forests are cleared is essential to maintaining their carbon-storing capacity and the ecological services they provide. They store large quantities of carbon as they grow, regulate the climate by pumping moisture into the air, provide watershed protection, help prevent soil erosion, provide a buffer against flooding, provide sources of food and habitat for an enormous range of species, provide fuel and construction materials and can even act as a pharmaceutical resource.

Biodiversity has been called 'the natural wealth of the Earth' and it is vital that we safeguard it. The Millenium Ecosystem Assessment was a four-year study of the world's ecosystems undertaken by 1,360 scientists, which concluded in 2005. It stated that 15 of 24 primary ecosystem services are being degraded or pushed beyond their limits. The data suggests that the rate of extinctions of species is now running at 1,000 times the rate that would be considered normal, based on analysis of the extinctions we can track through the fossil record. This depletion of biodiversity results in further degradation of the ecosystem services upon which human life depends. Ultimately, this is what makes climate change so important to us as individuals: because biodiversity and ecosystems are vital to supporting human life.

'Biodiversity has been called 'the natural wealth of the Earth' and it is vital that we safeguard it.'

Industrial exploitation of tropical and subtropical rural areas has often increased poverty in the developing world. The destruction of old growth forests, watershed loss and desertification have damaged or destroyed the habitat of many of the world's most endangered species and vulnerable people. Logging, shifting agriculture, population growth and the oil and mining industries are all putting more pressure on global forestry and without some form of policy intervention, rates of deforestation are likely to increase.

Climate change is exacerbating the dangers that biodiversity loss poses. As things stand, the cost of these biodiversity losses fall most heavily on the poor of developing nations, as many of them have livelihoods that are largely dependent on land use. The 2008 TEEB report showed that the true global cost of forest loss alone is anywhere between $2 to $5 trillion every year – a far greater sum than that lost in the financial crisis, said in mid-2008 to be around $1.5 trillion. Given the tension between developed and developing nations on how to fund adaptation and mitigation in the developing world and how to encourage the developing world to agree to binding targets, the inclusion of forestry in the carbon markets could be a solution.

REDD to the rescue?

The political debate gained new momentum in late 2005 when a proposal submitted to the UNFCCC by Papua New Guinea and Costa Rica called for the inclusion of avoided deforestation into any future climate regime. It suggested a new system of credits, based on the performance of entire countries in reducing the loss of their native forests, providing a financial incentive for the protection of existing forests. This gained pace at the conference of the parties to the UNFCCC in November 2006, when it was agreed to explore a proposal by Brazil that would provide incentives to reduce deforestation emissions in developing countries.

At the 2007 Bali meeting, an important step was taken towards reducing emissions from deforestation with an agreement to launch a framework for demonstration activities under Reduced Emissions for Degradation and Deforestation (REDD). This was intended to allow different approaches to reducing deforestation and forest degradation to be tested in preparation for covering these issues in a post-2012 agreement. The demonstration activities are supported by the World Bank's Forest Carbon Partnership Facility, also launched at Bali.

The consensus was that some form of mechanism to reward avoided deforestation would help balance the inequality between developed and developing nations in addressing global GHG emissions, as the majority of forest is found in sovereign territories of developing economies, for example in Latin America and Africa. These nations are more likely to be most at risk from the consequences of dramatic climate change and in need of funds to help adapt to changes in the climate.

Barry Gardiner, chairman of the GLOBE Forestry Dialogue, has warned that the current REDD framework actually provides a perverse incentive to cut down forests. The idea is that we should reward developing countries that slow down their rate of deforestation; paying, in effect, for the difference between what they have cut down and what they might otherwise have cut down. The method is one that calculates a historical rate of deforestation and then extrapolates it into the future gradually reducing it over decades. Any improvement between this 'reference rate' and the country's actual rate of deforestation is then measured as 'avoided deforestation' against which tradable credits may be issued.

Technically avoided deforestation is a calculation that is measured as a deviation from a hypothetical. Any indication by the international community that it will adopt a system of payment for avoided deforestation automatically encourages developing forest nations to increase their current destruction of forest. By doing so they will increase their ultimately agreed reference rate and maximize the differential between their hypothetical loss and the actual loss they incur. By artificially boosting the measurement of avoided deforestation in this way, countries would be able to secure additional payments for actions they would have taken in any event. To compound this problem; countries whose historic rate of forest loss is commendably low, may feel themselves aggrieved by a system that ensures larger payments to those who have failed to preserve their forest cover whilst countries with more successful conservation records benefit least of all.

Another concern is the potential is that rich nations could buy up significant tracks of land to gain access to the carbon opportunity. Growing numbers of developed economy investors are already buying land in fertile areas of developing economies. Such opportunistic purchases could lead to circumvention of the rights of the people and places that such projects are intended to assist. Much marginal or forested land in the developing world is common property, which means that local communities could fail to benefit from such arrangements.

Valuing forests

There is a clear economic relationship between the existence of the forest and the function of the economy that goes far beyond the concept of an environmental good. Fortunately, stopping their destruction can be done comparatively quickly and cheaply.

In a 2008 report published by Policy Exchange and entitled *The Root of the Matter: Carbon Sequestration in Forests and Peatlands*, Ben Caldecott argues preventing deforestation and stopping peatland destruction are some of the cheapest and most effective ways of reducing global GHG emissions. These methods of reducing emissions are dramatically cheaper than all other mitigation options available – as low as $0.1 per tonne of CO_2.

Tropical forests are divided by scientists into four main groups – the Amazon, the Guyana shield, the Congo basin and the Southeast Asian forest. The Brazilian Amazon alone, which is said to store some 60–70 billion tonnes of carbon, discharges 55% of the world's freshwater and generates 20 billion tonnes of rain per day, supplying water to the $1 trillion agricultural industry in the La Plata river basin in Argentina. While scientists don't yet know what would happen to rainfall if the Amazon is razed, replacing Brazil's hydro capacity – which accounts for 70–80% of its electricity generation – would cost $100 billion. This figure implies a value of $260 per ha for Brazil's remaining forests, even if we only value it in terms of water management.

Action is being taken on protecting forestry, despite it remaining outside existing international climate change agreements. Both private equity houses and countries have begun to address the need to protect such a resource.

Case study: Congo Basin Forest Fund (CBFF)

The £50 million Congo Basin Forest Fund is a multi-donor country fund established to take early action to protect the forests in the Congo Basin region. Covering 200-million hectares and including approximately one-fifth of the world's remaining closed canopy tropical forest, the Congo Basin Forests are also a very significant carbon store with a vital role in regulating the regional climate; and harbouring diversity of global importance. The Fund is intended to support transformative and innovative proposals that will help the people and institutions of the Congo Basin to manage their forests; helping local communities find livelihoods that are consistent with forest conservation and developing new approaches, which will bring genuine change and ensure future sustainable forest management; reducing the rate of deforestation.

The Fund will provide a source of accessible funding and encourage governments, civil society, NGOs and the private sector to work together. The CBFF is initially being financed by a grant of £100 million from the British and Norwegian Governments. It will be managed and disbursed by a Secretariat based at the African Development Bank (AfDB). Although the Fund is intended to directly benefit the countries of the Congo Basin, it is also considered as a global public good, which will have continental as well as global benefits. The forests of the Congo Basin, the second largest rainforest in the world, are an essential resource providing food, shelter and livelihoods for over 50 million people.

Whether we choose to include forestry under a new post-Kyoto agreement, one thing is clear: a failure to incentivize the protection of global forestry, could result in our being incapable of preventing dramatic climate change.

The International Policy Framework | 5

It is clear that little to no action in addressing climate change will take place without certain things: a strict and diminishing carbon cap, a carbon price, regulatory support for new cleaner technologies and the political will to implement the change to a low carbon economy.

In order to understand where we are in addressing climate change, it's important to know what's happening on the international stage. The Kyoto Protocol, which was adopted in 1997 and came into force in 2005, is an agreement between nations to address the causes and impacts of climate change in order to keep the global climate balanced. It is intended to provide a means of cutting the emissions of of six primary GHGs: carbon dioxide, methane, nitrous oxide, hydrofluorocarbons, perflurocarbons and sulfur hexafluoride.

The international community is currently in the process of negotiating a post-Kyoto agreement, in the hopes of creating a treaty which will address global GHG levels, while encouraging economic growth decoupled from emissions growth. The process is being hampered by the tension between the developed and developing worlds, with regard to the financing of climate change mitigation and adaptation, and the issue of binding emissions targets for the developing world.

The Kyoto Protocol

Kyoto was the first international legally binding framework for cutting GHG emissions. In recognition of the fact that the developed world is responsible for the majority of GHG emissions to date, through over 150 years of fossil-fuel consuming industrial activity, the Protocol enshrined the concept of 'common but differentiated responsibilities'. This has meant that the larger burden of the mitigation of GHGs and possibly more importantly, investment in the mitigation of GHGs, has been seen as the responsibility of the developed world.

Developed Nations	Annex 1	Fixed emissions caps	Examples: EU nations, Canada, Japan, Russia
Developing Nations	Non-Annex 1	No emissions cap	Examples: Brazil, Mexico, South Africa

The treaty set out a series of different targets for signatory countries, and each country is entitled to meet those targets through whatever measures they deemed appropriate. In order to help countries meet their targets in the most cost-effective fashion, however, the treaty included 'flexibility mechanisms' – methods such as emissions trading, the Clean Development Mechanism (CDM) and Joint Implementation (JI).

Kyoto and the CDM

The decarbonization of industry is a requisite if we are to successfully tackle climate change, which means that there is a vital role to play for emissions offsetting under the CDM. The thinking behind the CDM is simple – offsets are economically efficient, therefore cheaper, which should cut global emissions (as low carbon projects are developed instead of high carbon projects) and provides benefits for developing economies as finance flows towards projects in the region.

In order to qualify to sell credits, each CDM project has to register with the UN's CDM Board. This regulatory body defines whether the project qualifies for credits, and has an audit and monitoring system in place, agreed upon at the international level. Under the Kyoto system, individual nations maintain their own registries of credits, which can be transferred under the International Transaction Log (ITL). This was set up by the UN to verify and register credits, track where they've been transferred and, critically, note their retirement – which means they can't be used again. This notion of offsetting emissions in one place by avoided emissions in another means that the CDM effectively works as an industrial subsidy from the developed world, in order to incentivize the transition towards cleaner energy.

The first objective of the CDM was to create a tradeable credit that would enhance the development of the carbon finance market and this is steadily being realized as the market grows more mature. The second objective, a contribution to sustainable development through the underlying projects has, to date, been relatively marginalized. In fact, there is a growing concern that it has failed to support such sustainable development.

In the negotiations prior to the creation of the Kyoto agreement, the CDM was originally intended to be a fund; it was even for a time called the 'clean development fund', as proposed by Brazil. Its purpose was to enable the flow of capital and technology to developing countries to enable them to provide clean energy infrastructure for local sustainable development. There is, therefore, an expectation among policy makers, governments and NGOs that this is what the CDM should deliver. Negotiations led to the transformation of the clean development fund into a market mechanism. That meant that the qualification criteria for projects under the new system became harder to meet. This makes them more expensive, and creates an environment where the financial benefit of a project can be deemed more important than its local benefit.

One clear example of the problem is the exclusion of forestry and small scale projects from the final agreement. According to

Lionel Fretz, chief executive of Carbon Capital Markets, in a 2007 letter to the Financial Times, 'Forestry, although clearly mentioned by the protocol and also a sector with definite sustainable development benefits, was kicked into the long grass. The critics' arguments are perfectly valid; namely that it is impossible to ensure that by preserving one area, loggers do not move elsewhere and that the forest can burn down, re-releasing CO_2. As a result, we have a subset of the carbon market that does not fit within the strict parameters of Kyoto, but which has been taken up by the voluntary offset sector. The same applies to smaller scale, sometimes village level, activities where the cost of compliance with Kyoto (certifiers, validators, consultants, etc) is prohibitive'.

This has resulted in a system with many critics. Many of the initial CDM projects were based on the removal of HFC–23 from industrial installations in China, a GHG with a global warming potential 11,700 times that of CO_2. These emissions were predominantly a byproduct of the manufacture of fridges and air-conditioning units and there was some question as to whether or not these gases would have been cut anyway under China's own anti-pollution laws. Critics complained that income from the sale of HFC–23 related Certified Emission Reductions (CERs) was, in fact, subsidizing the manufacture of HCFC–22, as a gas that was being used instead and so undermining the use of alternative refrigerants with lower global warming potentials.

Another problem is that projects with increased efficiency can simply mean that more electricity is generated; resulting in no difference to the amount of GHGs generated, or sometimes leading to increasing levels of emissions. This can happen when the increased profitability of the generating company might yield additional funds to invest in new generating capacity, so raising GHG emissions.

At the same time, the great majority of CDM projects under way are located in India, Brazil, China and Mexico – all countries in which GHG emissions are rising rapidly as a result of rapid industrialization, much of it caused by the relocation of manufacturing industry from Annex 1 countries. Critics have claimed that this means that the issue of whether actual emissions reductions have been achieved is open to question. It also puts added pressure on the question of when these industrializing nations should take on their own absolute and binding emissions targets, a bone of contention in post-Kyoto negotiations.

As a mechanism for the funding of mitigation in the most vulnerable areas of the developing world, CDM seems to have failed. The most vulnerable countries are often those with the lowest available funds to address the problems. Although CDM is responsible for directing investment and technology transfer to developing economies, over 70% of investment has been in China, which may be developing, but is also one of the world's largest economies. By October 2008, 1191 projects had been registered (with over 2,000 more in the pipeline), over 202 billion CERs have been issued and 1,330 billion CERs are expected until the end of 2012. Yet there are only around 25 CDM projects in Africa – a continent that will be seriously affected by the impacts of climate change.

On a more positive note, however, there has been a significant growth in the number of clean energy projects being implemented in the developing world. We are now seeing a situation where the majority of CDM projects are related to the generation of renewable power: in 2007, clean energy projects accounted for 64% of CERs generated. This was an increase from 33% in 2006 and only 14% in 2005.

There is no question that the CDM needs to be reformed. This is vital to ensure the environmental integrity of the system and the acceptance of carbon trading as a whole, as a cost effective way to combat climate change. The fairly robust system for verification and certification under the Protocol should be encouraged and expanded, but possibly with a refocus upon its underlying precepts. One requirement should be that any carbon reduction should also have a net benefit for the environment, another that sustainable development should be encouraged or even prescribed. A further goal should be the expansion of these regulated standards to cover non-regulated offsets.

The difficulty is that the process of quantifying emissions, let alone emissions reductions, is an immensely complex task. In the end, while the underlying philosophy behind offsets needs clarification, buyers need to be certain that they bring the benefits they expect. And that means globally accepted standards and methodologies are required.

When the CDM programme began there was little clarity and no long-term investment framework and this remains the case prior to a post-Kyoto agreement. One thing that tends to be forgotten is that investors are doing a job and their job is to make money. It is not to take risks and support new initiatives; it *is* to minimize risk or *take* a risk in the hopes of a great reward.

Many of today's complaints about the CDM come from those who are watching the outcome of a few short years of implementation, of a process created from scratch to appeal to the majority and in a political environment fundamentally different from today's. This system was created long before concern about climate change and how it should be addressed became a mainstream issue. That means that the knowledge we have gained can be used to improve the system.

Additionality

A critical issue for CDM is 'additionality' – that is, whether the project issuing the credits would have existed under a business-as-usual scenario and whether it would have been financially viable without the financial incentives provided through the CDM or other Kyoto mechanisms.

For a project to be eligible to sell offsets, it is supposed to prove that it is additional. Judging additionality has turned out to be difficult – the CDM board itself is reviewing its own procedures on this point. It is impossible to prove a negative – that a project owner would not have built a project, or changed to a cleaner fuel, over the future lifetime of a project.

It might be more practical if we recognized the CDM specifically as an international subsidy for the deployment of renewable energy and energy-efficiency projects. The key issue is to ensure that high-emissions developments are replaced by those with low emissions.

The reason that this has become such an issue is a growing concern that the use of offsets simply displaces emissions, rather than cutting them. Given the use of CDM credits within the European carbon-trading scheme, there have been growing fears that the industrialized world is simply outsourcing its emissions to the developing world. CERs have traditionally been cheaper to generate than EU credits (an EU ETS credit on average cost $24.30 in 2007 compared with around $13.50 for a CER, according to the World Bank's *State and Trends of the Carbon Market* report).

The reason that it is important to ensure that credits are additional is to ensure that they are valid for use with a cap-and-trade scheme. The first, and still the largest, carbon trading scheme is the EU's Emission Trading Scheme.

EU emissions trading scheme – cap-and-trade in action

Even before the Kyoto Protocol was fully ratified, the EU set up a market to control carbon emissions and enable the region to meet the targets that it had agreed under Kyoto. It is by far the largest and most comprehensive market action to date, and plays a critical role in the EU's climate policy, currently covering over 40% of the region's GHG emissions. Unquestionably mistakes have been made as it has developed, but important lessons have also been learned.

The EU ETS was structured in three phases: a first trial phase from 2005 to 2007; a second phase corresponding with the Kyoto Protocol commitment period from 2008 to 2012; and a third phase running between 2013 and 2020. The amendments to the EU ETS made in 2008 will come into effect in January 2013.

The scheme was intended to penalize those who failed to address their emissions, and provide a revenue stream for those who, ostensibly, had invested in emissions reductions. The idea was that as the targets shrink over time, the supply of credits or European Union Allowances (EUAs) would become smaller, making it more expensive for emitters to buy their way out of trouble. This was intended to stimulate investment in low carbon alternatives to traditional business practice. At the same time, through the purchase and retirement of credits issued under the EU ETS, purchasers would be contributing to the overall reduction of carbon emissions within the EU.

Each Member State of the EU was given a target for emissions cuts. They then set National Allocation Plans (NAPs) and allocated the relevant number of European Union Allowances (EUAs) to match that target. One EUA represents one tonne of CO_2 equivalent (CO_2e). Companies regulated under the scheme have to surrender sufficient allowances to equate to the number of units of CO_2e that they have emitted that year. Any shortfall can be covered by purchases of allowances on the open market.

Any company that fails to match emissions levels to their allowances is fined (at a rate of €100 per tonne in 2008 or around £90) and is also required to meet the shortfall through the purchase of further credits. Under the original provisions of the EU ETS, carbon credits, including CERs, created from emission reduction projects in developing countries, can be traded and used interchangeably with EUAs.

The trading scheme covers over 7,300 companies, relating to around 12,000 installations. These are collectively responsible for about half the EU's CO_2 emissions and around 40% of its total GHG emissions. It addressed five key areas associated with high GHG emissions: power generation, oil refining, mineral industries (such as cement, glass and ceramics), steel and other metals, pulp and paper and other large combustion plants. It didn't cover transport (responsible for about 21% of EU emissions), buildings emissions from houses and small businesses (at around 17%), or agriculture (at around 10%).

The EU ETS was heavily criticized following its first phase, especially with reference to the decision to gift companies with their carbon credits, as well as the mismanagement of projections of the amount of CO_2e that would be emitted during the first phase. The combination of the two meant that not only were industries given a financial windfall through free carbon credits, but the demand for EUAs was so low that at one point the price of a tonne of carbon fell to less than a euro.

Carbon credits that were given to industry were intended to offset the cost of the development of alternatives to fossil-fuel based industrial practices. This didn't take place on any significant level. Generators for example, benefited from high electricity prices reflecting the cost of carbon emissions, but received the carbon credits for free. Renewable energy sources did not see an equal investment and deployment. A study for the WWF by leading carbon market analysts Point Carbon showed that the windfall to power generators in the UK, Germany, Spain, Italy and Poland alone could be as high as €71 billion over the 2008–2012 phase of the EU ETS.

Allocations of carbon credits have become tighter since then, with the market expected to be short of credits overall during the second phase of the scheme, although its unclear what impact the financial collapse is likely to have on demand. Certainly the early signs are that falls in industrial output are depressing demand for credits, leading to volatile carbon prices. This can present a problem, especially given that it is the carbon price that is supposed to provide incentive for long-term investment in alternatives.

In January 2008, the EU announced details of its Climate Change and Energy Bill. This outlined its own targets of cutting CO_2e by 20%, generating 20% of its power from renewable energy, and increasing efficiency by 20%, all by 2020. The Bill included a reform of the EU ETS scheme that would see credits allocated to industry through an auction, rather than freely allocating them under NAPs.

While member states were entitled to auction up to 10% of their allowances in Phase II of the EU ETS (2008–12), in practice most have optioned to auction only around 4%. The power sector – which generates the majority of EU emissions – will face full auctioning from the start of the new regime in 2013. Oil refineries and airlines, as well as some other industrial sectors, will have to purchase 20% of their credits, rising to 100% by 2020 although the Commission has said exceptions may be made for sectors vulnerable to competition from producers in countries without comparable carbon constraints.

In addition, auctions will be open: any EU operator will be able to buy allowances in any member state. It will mean allocations set at the EU level, as well as an increase in installations covered and will cut allowances year on year to cut emissions levels by 21% from 2005 levels. In sectors not covered by the ETS such as buildings, transport, agriculture and waste, the EU wants member states to reduce emissions to 10% below 2005 levels by 2020.

While auctioning will provide an additional incentive to invest in low carbon technologies, there is no guarantee that the proceeds of such auctions will be used to help fund the development and deployment of such technologies. The Commission has estimated that the revenues from the auctioning could amount to €50 billion annually by 2020. Revenues resulting from the ETS will accrue to member states and it is suggested that revenues should be used to help the EU to adjust to an environmently friendly economy by supporting innovation in areas such as renewables and carbon capture and storage. It is unclear if, or how, this will be achieved however. For example, the UK government has so far refused to commit to using funds raised in this manner to be invested solely in renewable energy and low carbon technologies.

Industry has been clear about its opposition to increasing stringency in the regime. The concern is that the cost of doing business within the EU could become uneconomic, as entire industries transfer their operations to regions with less stringent carbon controls. In 2008, Jeroen van der Veer, chief executive of oil multinational Shell, warned that the auctioning of carbon allowances could mean that Shell would not invest further in Europe.

The question is how far such warnings by industry should be of concern. Naturally, it is in the best financial interest of energy-intensive industry to lobby against additional costs. In 2008, Green Member of European Parliament Claude Turmes said that claims about potential 'carbon leakage' by industry had been seriously overstated. He said: 'According to the energy-intensive industry lobby, EU industry is heavily exposed to global competition. But exposure to non-EU competition is not even 2% for the EU's lime and cement industry and around 5% for EU refineries. For the steel sector competition does not reach 20%.'

If the EU is to achieve its goals, weak legislation and unclear long term policy signals help no-one – in many ways it is the uncertainty of the impact of carbon pricing that is of most concern. The most important factor in developing new low carbon alternatives to current power and manufacturing processes is a strong and consistent policy environment. And that means that if the EU wants to reach its proposed targets, hard choices are going to have to be made.

Voluntary vs compliance markets

One of the challenges for today's carbon markets is to understand the role of the voluntary carbon offset markets in the overall carbon mitigation picture. The idea of voluntary carbon offsetting is straightforward. It allows those individuals or groups who pollute to reduce the impact of that pollution by paying another individual or group not to pollute. For example, a flight to the US might produce a set number of tonnes of CO_2e emissions and, in order to offset this, an individual might pay a carbon-offset company to plant some trees, or convert lights in a developing country from gas to solar power.

Environmental groups have raised concerns over the value of offsetting, pointing out that it does not actually remove the CO_2 released into the atmosphere by activities such as air travel and driving cars. The perception has become that polluters are simply paying others to salve their climate change conscience, as they continue to emit CO_2.

Initially, there was little clarity about the reliability and probity of some of the offsets claimed in the voluntary sector and an increasing media spotlight found faults in the system. This has had an increasingly negative impact on public perception of the carbon markets. With the regulated market for carbon credits expected to reach over \$68 billion by 2010 and revenues for the unregulated voluntary sector rising to around \$4 billion in the same period: it is becoming ever more important that the market for carbon credits as a whole is seen as beyond reproach.

A number of standards have been launched to ensure that voluntary carbon offset credits are 'high quality'. A leading example is the Gold Standard, an NGO initiative that provides tools to develop emission-reduction projects that result in real and additional emission reductions, promote the transition to sustainable energy systems and secure both local and global sustainable development benefits. The International Standardization Organization (ISO) in Switzerland has also provided a GHG standard to verify the quality of emissions reductions. There are a range of other standards, focused on certifying sellers, products and services, as well as assessing claims of corporate carbon neutrality.

Many commentators applaud the various yardsticks, saying they provide effective ways to guarantee that carbon reductions are delivered and not double-sold and, clearly, there is an acceptance within the market that standards are necessary. But are 20 different standards helpful? There are already globally accepted standards for the compliance markets and for CERs, which are verified by the CDM Executive Board to ensure transparency and integrity.

Should all offsets be regulated?

In 2008, the UK's DEFRA published best practice guidelines suggesting that carbon credits for the voluntary market should be sourced from established Kyoto markets. It suggested that the only credible form of offset was the purchase of a compliance market credit and its cancellation. This position was based on the standards and transparent audit trail imposed on project developers within the Kyoto compliance market to ensure emissions reductions actually take place.

This move was heavily criticized by the voluntary carbon credit industry, who argue that regulation of the voluntary markets would increase transaction costs and keep many innovative and sustainable projects out of the markets. One of the great benefits of the voluntary markets is their ability to address smaller projects, or bundle a series of projects together, in a way that is non-commercial under the existing CDM process. Yet it is difficult to see how else to move forward if the carbon markets are to gain global integrity.

Both companies and consumers need to feel certain that the offsets they're purchasing actively stop GHG being emitted. Buyers of regulated credits must be sure their assets have value and are sourced from projects which make a genuine contribution to a long-term climate solution. By supplying credits only from this regulated market, buyers won't have to brave the range of standards that are already on offer as the basic rule could be 'if its UN approved, you don't have to think about it'.

Critics have claimed that solely using CERS and/or EUAs could decimate the developing voluntary market and that many verified emission reduction (VER) credits do meet a range of market standards. The fear is that sticking to those credits regulated in the compliance markets could mean the exclusion of legitimate and effective carbon emission reduction projects.

While there are valid concerns about the impact of using compliance credits in the voluntary markets, there is no question that it will have a direct impact on the level of offsets available. This is surely a positive move, as lower credits available for trade will increase pressure on companies to invest in lowering their emissions profile. If this approach is taken up, there may be a knock-on effect on the VER market but, as the credits generated will remain cheaper than those generated from the compliance markets, it's likely that an appetite will remain. The key difference that increasing interest in the purchase of compliance credits from the voluntary sector will have, is increased prices for compliance credits and increased pressure to lower emissions in those industries monitored under the EU ETS.

The trouble with Kyoto

While the introduction of the Kyoto Protocol was unquestionably a key step for the international community in joining forces to combat the effects of climate change that doesn't mean that it has been entirely successful. Critical to the success of any compliance regime is its governance and how it is policed. Under the Protocol, nations that fail to meet their emission targets are required to make up the shortfall, plus an additional 30%, over the next emissions target period. A financial penalty may also be assessed on violating nations by suspending their eligibility to sell emission credits under the Protocol's emission trading system. Of course, we won't know how effective this is going to be until we enter the post-2012 period.

There are still unanswered questions about the implementation of the Treaty. Might it not be easier for countries that have not achieved their emissions targets to opt-out of the system altogether? Why would they want to be reinstated? It seems unlikely that a country that has failed to meet its initial targets is going to be able to meet an increased target of 30%. There are alternatives that might be considered – if one country under the agreement fails to meet its targets but another country exceeds its targets, perhaps the UN could mandate that the non-compliant country must buy those excess credits. This would not only provide an economic disincentive to fail in cutting emissions, but would also provide a strong economic incentive to overshoot the target. As things stand, it is unclear what, if any, consequences will be enforced, which does create a problem. Enforceability is a key issue and must be resolved in any post-Kyoto climate change agreement if it's going to have any real teeth.

Whatever the concerns, the Protocol was a vital first step towards a truly global emissions-reduction regime. For example, by reducing their own emissions first, developed nations might gain the legitimacy required to convince developing nations also to reduce their emissions. The growing number of regions that are developing their own cap-and-trade schemes, including Australia, New Zealand, Canada, Japan and the US, suggest that this may well be the case.

Much of the criticism around the Kyoto Protocol is over political realities and the limitations of the treaty, especially with regard to the potential economic consequences. Critics say a 5% cut in emissions to 2012 will accomplish little, especially without the US on board and there are strong doubts about the ability of many signatories to meet their targets. The latest scientific research suggests that we need to cut emissions by 60–80% by 2050 to have any serious change of preventing serious climate change – that's a dramatic shift in goals over 40 years. At the very least we need to set interim goals and outline how we plan to address failure to meet those goals.

The potential economic impact of the Protocol on participating nations is another critical issue as they work towards meeting their emission targets. Many economic sectors may be disrupted as companies are forced to introduce new technologies and procedures to reduce their GHG emissions. In some cases, so the argument goes, the cost of introducing these technologies and techniques may be such that some companies will either close, or move to a non-compliant country.

Growing concern about the lack of binding targets on developing economies such as India, China and Brazil is also causing problems. While China admitted in 2008 that its emissions were now on par with the US, according to some methodologies China has been emitting more than the US since 2007.

One of the biggest problems with the Kyoto Protocol, and with negotiations for a post-Kyoto agreement, is the need to structure an equitable agreement.

Reaching an equitable agreement

Ensuring that any international climate change agreement is equitable is a central requirement. The richest countries in the world are likely to be least affected by climate change, while the impact is likely to be hardest in the developing world. Yet the majority of GHG emissions have predominantly been generated by the actions and economies of the developed world.

Many developing economies have been adamant that they won't accept binding cuts under any international agreement until they have managed to pull their poorest members out of poverty. Others see the discrepancies between binding reductions for some countries and not for others as unfair, especially as the emissions of some leading non-Annex 1 countries (such as China and India) begin to rival those of Annex 1. Whatever your starting position, a critical issue is fairness: why should one party suffer more than another?

There have been many suggestions on how to make the system more equitable. One such is calculating emissions per capita and setting emissions caps per country on that basis, which theoretically sounds fair. It doesn't, however, address the concerns of the developed economies – that they will have to fund mitigation in the developing world while those economies can grow with impunity. At the heart of the original Framework Convention was that countries should have shared but differentiated responsibilities. The question remains what is the best way to share that responsibility fairly?

The CDM was intended to make the process of addressing emissions more equal. Not only was it intended to ensure that cuts in emissions growth took place at the cheapest price possible, it was intended to act as a means of transferring funds and clean technology from developed to developing economies, through projects that were also supposed to have positive social benefits.

The purchase of credits, along with investment in the carbon market, is not just helping western companies meet their compliance quotas or salve their consciences. It also acts as an incentive for greater investment into new clean technologies, generating huge flows of global finance from rich to poor, allowing for technology transfer between countries and enabling wider sustainable social and economic development in developing parts of the world, while actively reducing emissions of GHGs.

A central premise of the CDM was to develop clean energy infrastructure for developing nations. GHGs are damaging on a global scale, no matter where they are emitted, so the idea was that the best way to start is by cutting them where it's cheapest – in the developing world. Technology has always been key to any emissions reduction project, whether building a wind farm for renewable electricity, or improving the energy efficiency of an industrial process. Technology transfer involves conveying of equipment, knowledge, operating skills and project management expertise from where it's developed to where it's needed.

This flow of finance and technologies also allows for far wider social and environmental benefits across the developing world, creating job opportunities and helping retrain the local workforce.

As carbon markets develop and more operators enter the market these social and human development aspects become ever more important. Buyers in the compliance and the voluntary sector will want to know they are buying 'premium' carbon from reputable sources that result in real, additional, emission reductions, promote the transition to sustainable energy systems and secure both local and global sustainable development benefit. The competition between carbon-traders at the heart of the market system should act as a stimulus to encourage best practice, increasing the benefits for project developers, for local people and communities and for the planet itself.

If best practice is encouraged, it can offer a vision of how markets can bring multiple benefits and investment in technological, environmental and human capital.

Alternative approaches

There has been a range of proposals made on how to move forward from the Kyoto Protocol, including proposals from the G8, which has made action on climate change a central tenet since the 2005 Gleneagles meeting, and from the US-backed Major Economies Initiative (MEM), a discourse launched by the US to attempt to set binding goals agreed by all economies.

According to the MEM, each country should establish mid-term national targets and programmes that reflect their own mix of energy sources and future energy needs. The question is whether or not working outside the Kyoto framework is going to be a help or a hindrance to the goal of fairly rapidly achieving an effective international climate change agreement.

Cutting emissions by industry sector

One of the ideas that has gained a great deal of currency over the last couple of years is that of taking a sectoral rather than a national approach to the problem of GHG emissions, which is strongly supported by the Japanese and the EU, especially with regard to addressing aviation and shipping. Under this approach, no one country is given a particular competitive advantage.

One approach is that of the US-based Centre for Clean Air Policy (CCAP), that all economies (including those of major developing countries) would agree to a voluntary-sector-GHG-intensity target per unit of production, such as agreed caps on levels of GHG emissions per tonne of steel. In order to provide incentives to industry, developing economies would receive technology incentives in exchange. The original sectors proposed by CCAP were major energy and heavy industries (e.g. electricity, cement, steel, oil refining, pulp/paper and metals), which would ensure significant reductions in emissions at a global level. According to CCAP, the inclusion of the top 10 largest GHG-emitting developing countries in each sector would insure coverage of 80–90% of developing country GHG emissions in each of the selected sectors.

Crucially for industry in the developed world, the system would also address concerns about competition by encouraging the participation of all major concerns in a sector in both developed and developing countries. The approach is growing in popularity. In early 2009, the EU's chief climate negotiator Jos Delbeke proposed a 'global sectoral crediting mechanism' to set caps on certain sections of the global economy. The idea is that the larger developing economies, such as China, would agree to caps on its power and cement industries.

A primary driver behind the increase in the concept's popularity is increasing understanding of embedded carbon and the enormous difficulties involved in ascribing responsibility for emissions. If sectors were regulated, with emissions capped and analysed, it would provide useful additional information to the supply chain debate. A primary example of the import–export problem is China and the amount of GHGs that its manufacturing sector emits in the creation of cheap goods that are consumed in the developed world. Without Western consumption, these emissions might not be taking place.

The sectoral approach does not cover all the bases. There are countries that have high GHG emissions for other reasons than industrial activity. Those countries with substantial oil or coal resources tend to have high emissions associated with their extraction and distribution, even if the majority of that fossil fuel is not consumed within the country itself. There are issues to be overcome in terms of the feasibility of measuring, auditing and accounting for carbon within the supply chain, but the launch of the PAS 2050 (a UK standard for carbon measurement discussed in Chapter 3) shows that carbon measurement can be achieved and also underscores the absurdity of national caps in a global economic environment.

The sectoral approach is a bottom-up approach to the problem of carbon emissions, as it is an attempt to get industry to effect change, rather than have it imposed from above. One key benefit is that it would address the issue of carbon leakage, as it would remove any incentive for a company to move to a region or country with looser carbon regulation.

A sectoral approach would require a strong regulatory body, able to cross borders and enforce compliance but this is likely to be a requirement of any future approach.

Contraction and convergence

One of the key issues underlying all post-Kyoto debate is how to make any international approach equitable. In 1990, Aubrey Meyer, at the Global Commons Institute, proposed the original idea of contraction and convergence as a means of achieving this. The concept was adopted during the original Kyoto discussions by India and in 1997 by the Africa Group of Nations. However, it never made it through the final Kyoto negotiations. The central concept of Meyer's proposal is that all GHG emissions should be capped at the level needed to prevent dangerous climate change within a framework that includes every country and that emission rights should be allocated to each country on a per capita basis.

Developing countries would be allowed to increase their per capita emissions up to a point, as developed countries brought theirs down – eventually converging at a point which would keep emissions at a target level. Meyer recommended that the ability to trade these entitlements would be on a regional not a country basis, such as the EU, the Africa Union and the United States. A gradually reducing global cap, built on global rights, resource conservation and sustainable system could be the best way to galvanize international action.

There has been severe resistance to this concept in the international community, predominantly because, while there may be a theoretical acceptance that no one individual has more right to emit carbon than another, the differentials between economies are now so vast that it would cause enormous problems to resolve the inequity. For example, the average emissions rate per capita in China is about 7 tonnes, in India it is about 4 tonnes, while in Europe it is around 12 tonnes and the US leads the way at nearly 22 tonnes per annum.

Similarly, as with the current Kyoto mechanisms, there would be a requirement for such emissions to be tradable, providing countries with low carbon emissions with an income stream. However, the process by which this mechanism would be managed, audited and policed is unclear. If it was possible to include a demand that all proceeds would be spent on mitigation and adaptation, this could prove more interesting but experience shows that such agreements can be difficult to achieve.

While most people would agree that the concept of a per capita allocation of carbon emissions permits is equitable, it is equally clear that no major developed economy is going to agree to such a deal in the short- to medium-term. If the US refused to ratify Kyoto, the likelihood of it agreeing to such an approach seems low. The greatest difficulty is finding a means of equitable change which is combined with a pragmatic approach to implementation within a short time frame.

Personal carbon tax

The idea of a personal carbon tax was originally conceived by David Fleming, and further developed by Richard Starkey at the Tyndall Centre; it has proved to have a great appeal to groups ranging from the environmental to the political. It advocates Domestic Tradeable Quotas (DTQs), a system under which individuals would be allocated a personal allowance of GHG emissions to use or trade.

The appeal of this concept is that individuals would be responsible for their own emissions. While the authors suggest that the only options are taxation or rationing, politicians, as a general rule, tend to prefer not to introduce obvious taxes. Personal carbon trading has appeal to both the political right-wing, focused on free markets and the political left-wing, focused on equitable distribution of rights. Despite the fact that lower-income households tend to spend a higher proportion of their income on power and heating, as a general rule, this group tends to emit lower levels of GHGs and, therefore, would be more likely to be sellers of credits – providing them with a revenue stream.

The benefits of personal carbon trading are obvious, as again the programme would drive behavioural change and the cost of carbon emissions would be clearly identified at every decision-making point. If personal carbon trading was part of a wider cap-and-trade scheme covering the whole economy, then individuals could potentially trade carbon allowances directly with businesses and other large organizations. From an economic perspective, this could help to identify the lowest cost emissions savings across the whole economy.

As ever, the fundamental problem remains the cost of implementing such a scheme. Not only would all goods and services have to have their embedded carbon calculated, but the full carbon lifecycle, including distribution and disposal, would have to be analysed. While the launch of the PAS 2050 may bring us one step closer to that goal, we're still some way away from having a detailed economy-wide understanding of where every GHG is emitted. Even if we achieve that, such a system would require a countrywide bureaucracy to manage each individual's carbon account. While in the UK that would fit in neatly with government plans to track all our personal information on ID cards, it sounds time-consuming, painful and the potential for error looks enormous, let alone the implications for personal privacy.

Achieving a post-Kyoto agreement

As we have seen, there are major issues to resolve if we are to achieve a post-Kyoto agreement, including burden sharing and funding. The issue of funding is contentious as the Framework Convention stipulates that the developed world must take the lead in mitigation, but it doesn't take into consideration the changing balance of emissions as developed economies progress.

The original Kyoto agreement includes explicit acceptance of the fact that developing countries would need financial assistance to cope with the impacts of climate change, and created the CDM Levy, a tax of 2% on the CERs issued through CDM project activity. This tax finances the Adaptation Fund, established to finance adaptation projects and programmes in developing countries particularly vulnerable to the impact of climate change. Yet the amount of funds raised to date are minimal at only a few million dollars.

The US, UK and Japan have led a funding programme involving 40 developed and developing economies that has resulted in about $6 billion of funding for the newly launched Climate Investment Funds, an amount expected to rise to $10 billion by 2012. Yet while the funds are expected to provide a major boost to cleantech in the developing world, it is still a relatively low figure for addressing the problem. Hundreds of billions of dollars are likely to be required to manage the consequences of flooded coastal regions, cities and agricultural land and the impact of drought on healthy soil and clean water supplies.

There is growing tension between the two groups (developing and developed world), both of which have seemingly reasonable perspectives. There is a demand that larger developing economies take on more of the burden of addressing climate change and, in turn, there is a demand from the developing world that action is taken to provide funds and transfer technology in order to help in the fight. Each group sees agreement on the other issue as the necessary first step.

What complicates matters even further is concern about the mechanism for transfering such funds, even if they were made available. The developed world wants to see a financial mechanism, with checks and balances on where the money's going. The developing world favours a fund which can be drawn upon as necessary. There are plans for a global effort to tackle climate change and poverty, spearheaded through the Strategic Climate Fund. Yet, developing country negotiators have expressed concern that the fund will be structured in such a way that funds are partly provided through loans.

Another area which must be resolved in a post-Kyoto deal is a system of governance which makes the operation of any international scheme both impartial and effective. The idea of giving responsibility to a body such as the International Atomic Energy Agency, which already has international enforcement powers, could be worth exploring.

There are positives about many suggested alternatives to Kyoto, and it is critical that we reach a resolution and begin the process of cutting carbon, lowering the probability of catastrophic climate change, by 2025. Therefore, our priority must be reaching a global agreement quickly and taking action as soon as possible.

There are clear flaws with the EU ETS, the CDM, current carbon targets and current financing methodologies, but it is likely to be easier to amend systems already in place than reach entirely new agreements. Kyoto took nearly 10 years to move from agreement to implementation and there is significant momentum for a post-Kyoto agreement in Copenhagen in December 2009. While there are significant barriers to overcome in terms of the details of a post-Kyoto deal, it's likely to be far easier to move an existing framework forward than start again from scratch.

While the situation may look difficult, there is growing international pressure to reach an agreement. According to negotiators in 2005, it was considered impossible even to have a discussion about future commitments and even at the 2007 Bali meeting there was hesitation. Yet, in 2008, China, India, South Africa and Brazil announced climate change mitigation plans, an implicit acceptance of the need to plan for emissions control or cuts.

There is no doubt that the role of the US is critical in terms of successful negotiation but following the election of Barack Obama in November 2008, hopes are higher that some kind of resolution can be met in Copenhagen. The new US energy policy looks set to include $150 billion in funding for clean tech and low carbon technologies, funded by a national carbon cap-and-trade scheme. If the US and China are moving towards the same point, there is hope for an agreement.

'Following the election of US President Barack Obama in November 2008, hopes are higher that some kind of resolution can be met in Copenhagen in 2009.'

The real question should be: how can we improve Kyoto and the CDM? What are the big political decisions going to be – the levels of ambition in terms of targets, technology and finance? Of course, there's one more thing to keep in mind and that's if we don't suceed in Copenhagen it is not necessarily the end of the line.

While an agreement at an international level would potentially speed the implementation of the technologies and tools that are required to transform the global economy to a low carbon basis, there is still an enormous amount that can be done outside such an agreement on a national, corporate and consumer basis.

What We Can Do

If we're going to address rising carbon emissions inside or outside an international emissions treaty, we have to learn to manage carbon in our lives and businesses.

There's little question that a high, stable and global carbon price within an international policy framework would be the most effective means of encouraging action on climate change, but there are already many activities being undertaken by different groups outside the specific framework of an international treaty. These actions can be used as a blueprint for how change can be achieved.

We're seeing significant action among the four groups which have the power, influence, impact and funds to effect the transition to a low carbon economy: countries, cities, corporations and consumers. Climate change may be the central problem, but if fears about energy security, peak oil, high fuel prices and economic consquences and opportunities are what drive solutions, we can only be grateful for the perfect storm. We're going to take a closer look at what is currently being done outside the Kyoto framework by some countries, cities, corporations and consumers in order to make that transformation happen.

Countries

While there is a current climate agreement under Kyoto, not all countries have ratified it and, of those who have, not all have binding emissions targets. Every country is going to have a different role to play in an international agreement and a range of approaches have been required in cutting carbon. At a national level, issues of economic and energy security have tended to outweigh concerns about responsibility for mitigating climate change. Yet growing global concern has clearly had an effect, as even those countries without binding emissions targets within an international climate change agreement are taking action to cut carbon.

We're going to take a look at the UK, the US and China as examples of different domestic responses to the climate change issue..

United Kingdom

While the UK is a signatory to the Kyoto Protocol it has already committed itself to long term climate goals far beyond Kyoto, with the 2008 Climate Change Act (CCA) and a target of cuts in emissions of 80% below 1990 levels by 2050. As a Member State of the EU, it is committed to the EU goals of 20% renewable energy (although only 15% within the UK itself), improvements in energy efficiency by 20% and emissions of 20% by 2020.

It also has a history in addressing climate change issues, having introduced its first Climate Change Programme in 2000 and updated it in 2006. The UK's original aim was to reduce CO_2 emissions by 20% from 1990 levels by 2010 – far above its Kyoto obligation of 12.5%. While it is expected to fail to reach its own internal targets it looks fairly certain, possibly because it tried to overshoot its goal, that the country will reach its Kyoto target in 2010. There are lessons to be learned from the UK's experience in attempting to structure a framework for the reduction of emissions, not least of which is the danger of creating overly complicated frameworks.

The UK introduced its climate change levy (CCL) in 2001, as a replacement for the Fossil-fuel Levy. The CCL is a tax on the use of non-renewable energy by businesses and the public sector. Non-renewable energy includes electricity from the national grid, gas, coal and liquefied petroleum gas. In order to offset the tax, a company can buy a levy exemption certificate (LEC).

The aim of the CCL was to encourage energy efficiency and reductions in GHG emissions but has no impact on power generation, the domestic sector, small business, charities or fuel oils. Natural gas in Northern Ireland and good quality combined heat and power systems are also exempt. This all sounds good unless of course, your business is energy intensive, in which case it is possible to get a Climate Change Agreement (CCA) with Defra, which comes with an 80% discount on the CCL. While that does come with a proviso for improving energy efficiency and cutting carbon, the system does seem overly complicated.

The UK also implemented a quota-based incentive system, the Renewables Obligation, which meant that UK utilities were obligated to generate a specified amount of power from renewable sources. If they failed, they could either pay a penalty or buy Renewable Obligation Certificates (ROCs) from renewable power generators.

One of the complications surrounding the system in the UK is that every MW of renewable power generates three credits – a ROC, a LEC and a renewable energy guarantee of origin (REGO). This can lead to double-counting of credits and may have been a contributory factor in the slow growth of renewable generation in the UK.

The system means that generators can sell the power with the REGO as proof of renewable origin, the LEC to a company to allow them to use fossil fuel and use the ROC to meet government targets on generation. Ensuring the right policy framework and that different government departments communicate, is clearly needed here. The appropriate policy framework is vital in order to ensure that the market operates in a transparent fashion, and it is needed to generate investment interest, one of the the primary ways to encourage further development.

Feed-in tariffs (setting a payment rate for power generated from renewable sources which is higher than that for fossil-fuel generated power) have proved enormously successful in over 40 countries. They have succeeded in increasing market penetration, producing lower cost renewable power, stimulating and expanding local industries, creating employment and attracting massive amounts of investment. That being the case, it's hardly surprising to see that in its most recent climate change legislation, the UK government accepted that a feed-in tariff has got to be a good idea. The government amended the CCA to include a feed-in tariff for microgeneration up to 3MW, which means it could be used for hospitals and communities, not simply homes. It won't affect the Renewable Obligation (RO), as the majority of utilities affected by the RO operating generating plants are far about the 3MW threshold.

The UK's climate change bill also included aviation and shipping in its targets. While this should have resolved a major area of contention for climate change campaigners, it didn't. Gordon Brown's decision to approve the expansion of Heathrow airport has caused a furore, as some calculations show an expanded Heathrow could use 20% of the UK's target emission by 2020. Research from the Tyndall Centre for Climate Change Research has shown that if the aviation industry expands as predicted, the whole of the UK's carbon emissions allowance could be consumed by aviation alone by 2050.

The UK Government also announced that it will be providing £100 million for the development of an electrified transport network, while the government-funded Technology Strategy Board's Low Carbon Vehicles Innovation Platform will provide £30 million for electric and low carbon vehicles and it defines low carbon meaning vehicles with emissions below 50g of CO_2 per km.

But perhaps the most interesting thing has been the introduction of the Carbon Reduction Commitment (CRC), a UK-wide mandatory carbon-trading scheme, starting in 2010. It is intended to cover large non-energy intensive businesses and public sector organizations and is the first of its kind, in that it is going to include major retailers, banks, water companies, leisure groups, financial services, universities and, of course, local authorities. More specifically, it affects those industries not already covered by the EU ETS.

The CRC is intended to help cut a further 1.2m tonnes of carbon from the UK economy by 2020. It's a scheme covering companies whose consumption of electricity is greater than 6,000GWh per year or with a bill of more than £500,000 and is expected to affect roughly 5,000 organizations (although the government has already said this could be increased to cover around 20,000).

Companies will have to start buying carbon allowances to cover their carbon emissions and that will involve measuring and recording energy use and calculating CO_2 emissions. All energy other than transport fuels will be covered, such as electricity, gas, fuel and oil. During a planned introductory phase, due to start in April 2010, all allowances will be sold at a fixed price. From April 2013, allowances will be allocated through auctions with a diminishing number of credits available over time.

Although mandatory, the scheme will involve self-certification of emissions, backed up by auditing, rather than third-party verification. Once emission allowances are auctioned, the income from the auctions will be recycled back to participants by the means of an annual payment based on participants' average annual emissions since the start of the scheme. Each payment will constitute either a bonus or penalty according to the organization's position in a league table. Penalties for under-reporting are expected to rise to £75 per tonne CO_2 from 2013 onwards. Participants in this carbon scheme will also be able to purchase (but not sell) emission allowances from the EU ETS at a price that is the higher of the EU ETS price or the minimum CRC floor price.

While the CRC will be useful as a mechanism for forcing companies to cut energy consumption, increase efficiency and cut carbon, its most exciting aspect is that it is taking the carbon issue to mainstream industry. It will force large corporations to take a comprehensive look at their emissions. Responsibility for energy usage and carbon management activities are often split across any number of departments, from facilities management, operations, logistics, finance, environmental, production, CSR, H&S, sustainability and training, to name but a few. Cutting emissions within a company will require all these groups to work together, cutting energy use, buying cleaner energy, streamlining logistics and so on.

It could encourage major infrastructure changes, such as on-site energy generation systems and green data-centre technologies, as well as an examination of the scope for savings through systems integration and intelligent control systems. It should even encourage the increasing implementation of eco-design in goods and services, as well as a focus on minimising waste in production.

United States

The US famously failed to ratify the Kyoto Protocol, despite playing an integral part in the negotiations under President Bill Clinton. In fact, as a US negotiatior, former vice-president Al Gore was in part responsible for the introduction of market mechanisms into the agreement. As the world's largest economy and certainly at the time, the world's largest emitter, the Bush administration's refusal to ratify the agreement could have caused the entire agreement to fail and had a catastrophic impact on the long-term future of the climate. The US has been much attacked for this failure, but what many seem to forget are the efforts undertaken by various individual States, cities and many US corporations.

'What many seem to forget are the efforts undertaken by various individual States, cities and many US corporations.'

Despite a federal refusal to adopt Kyoto, many States have already committed to carbon cuts and legislation to support the shift towards a lower carbon economy. The following States have already made commitments to emissions cuts, which range from cutting emissions from 1990 levels by 2010 to 80% below 1990 levels by 2050: Arizona, California, Connecticut, Florida, Hawaii, Illinois, Maine, Massachusetts, Minnesota, New Hampshire, New Jersey, New

Mexico, New York, Oregon, Rhode Island, Utah, Vermont, Virginia and Washington.

A number of States have implemented climate change legislation and a large number of public utilities have renewable standards. Renewable Portfolio Standards (RPS), which demand that public utilities purchase a percentage of their power from renewable sources, or source credits from someone else in the marketplace, have been adopted by 26 states (plus the District of Columbia). According to analysis from the Lawrence Berkeley National Laboratory, half the total wind power installed between 2001 and 2006 resulted from State standards. The US federal government introduced the Production Tax Credit (PTC) in the 1990s, a tax benefit that has driven enormous growth in renewable energy deployment in the last few years in wind, solar, biofuels and qualified fuel-cells.

The election of 44th President of the United States, Barack Obama, to the White House in November 2008 has led to a significant change in the US position. Obama has pledged to cut CO_2 emissions in the US by 80% by 2050 (compared to 1990 levels). Among the policies that Obama has proposed are a windfall tax on energy companies to provide relief to US families, $150 billion investment in renewable energy and energy efficiency over the next 10 years, targets for achieving 10% of US from renewable sources by 2012 and 25% by 2025, strengthening of the national grid, a national low carbon fuel standard and 1 million Plug-In Hybrid Electric Vehicles (PHEV) on the road by 2012 and the launch of a federal cap-and-trade market to help cut those emissions. The package also includes a long term extension to the renewable Production Tax Credit. Other goals include eliminating oil imports from the Middle East and Venezuela within 10 years and making the US a global leader on climate change issues.

That doesn't necessarily mean that the US will agree on a post-Kyoto international climate change agreement, but it is a massive shift in its position. While there is significant momentum for the implementation of these goals, they may prove difficult to achieve.

Industrial lobbyists are likely to cause an outcry about increasing costs during a downturn, as utilities, car companies, oil and coal companies and utilities fight any new legislation. The new Democratic administration will need to deal with fallout from a global recession and with the US deficit expected to be near $1 trillion, action in the domestic market is going to be critical. The President will also need both congressional and senate support for any international treaty – it takes a two-thirds majority to ratify an international treaty.

Any improvement in US environmental performance is likely to continue to be driven by issues surrounding energy independence, as well as investment in clean technologies, as a means of boosting the economy. Clean energy, in particular, is proving to be a convenient policy tool, as it addresses four major issues that dominated the election campaign: energy security, the economy, employment and climate change. While the key agenda items in US politics are the Iraq war and the economy, Obama has made clear links between investment in the green agenda as driver of economic growth and recovery.

The issue of cap-and-trade will probably remain contentious. While the US sulphur market is the showcase of how such a market can cut pollutants while not imposing costs, industry is demonstrating significant opposition to such a market. On a more positive front, some of Obama's early appointments, including a Nobel Prize-winning scientist, have shown strong support for climate change action. He appointed Steven Chu, the director of the Lawrence Berkeley National Laboratory, as Energy Secretary, and Nancy Sutley, deputy mayor of Los Angeles for Energy and Environment, as head of the White House Council on Environmental Quality. Carol Browner, former head of the Environmental Protection Agency (EPA) under President Clinton, will be responsible for co-ordinating energy issues in the new administration. She also serves on the board of US commodities exchange APX, which provides the platform to track allowances in the Regional Greenhouse Gas Initiative (RGGI) in the northeastern US.

While the federal government has been slow, many States have already taken action, committing to carbon cuts and legislation to support the shift towards a lower carbon economy. The following States have already made commitments to emissions cuts, which range from cutting emissions from 1990 levels by 2010 to 80% below 1990 levels by 2050: Arizona, California, Connecticut, Florida, Hawaii, Illinois, Maine, Massachusetts, Minnesota, New Hampshire, New Jersey, New Mexico, New York, Oregon, Rhode Island, Utah, Vermont, Virginia and Washington.

The combination of a growing will to implement legislation at the federal market and the significant strides already made at state policy level make it clear that the US is heading in a new direction, whether it be inside or out of Kyoto. Each State has decided its own target and its own policy framework for reaching that target, much as the EU ETS was the European Union's response to its overall Kyoto targets. Understanding the different routes that the individual States can and have taken is critical to understanding how ready the US actually is to move forward on climate change.

Cap-and-trade among the states

Many states have formed alliances to introduce regional emissions trading with cap-and-trade schemes. The Climate Registry is a non-profit organization whose goal is to establish consistent, transparent standards throughout North America for businesses and governments to calculate, verify and publicly report their carbon footprints in a single, unified registry and it covers the majority of US states. There are 40 US states, 9 Canadian Provinces, 5 Mexican states and 3 Native American nations participating in The Climate Registry. The Registry publishes GHG emissions data following the standards of the World Resources Institute and World Business Council on Sustainability (WRI/WBCSD).

The first of the state cap-and-trade schemes, the Regional Greenhouse Gas Initiative (RGGI), launched in 2008 and covers Connecticut, Delaware, Maine, Maryland, Massachusetts, New

Hampshire, New Jersey, New York, Rhode Island and Vermont. It aims to cap emissions at 2009 levels and then cut them by 10% by 2019. It only covers electricity generators and has 2 compliance periods of 3 years, running between 2009 and 2014. The cap, set at 188 million tonnes of CO_2 per year, will shrink 2.5% annually between 2015 and 2018 for a 10% total reduction. All revenues from permit auctions are to be used to fund renewable energy and energy efficiency programmes in the ten RGGI states. The first auction was held in September 2008 and raised $38.5 million. By the third auction in March 2009, the funds raised were over $117 million.

The fact that the scheme began with an auction (rather than simply issuing credits for free, as was the case within the EU ETS) shows the lessons that have been learned in how to use the markets to limit carbon emissions, forcing polluters to pay for the privilege. Other issues, however, remain unresolved, such as whether or not the ratchet in the scheme is sufficient to have any significant impact on emissions; whether or not the oversupply of emissions credits in the early phases will mean the price remains too low to have an impact; whether or not it should cover significant industrial operations as well as electricity suppliers; and whether or not the funds raised at auction will be effectively used or sufficient to have significant impact on carbon reduction plans within the individual states.

The initial target was set roughly 4% above the annual average within the RGGI states between 2000 and 2004 but emissions have been lower since, due to a combination of warmer temperatures and a switch from coal to natural gas. This means that the price of carbon emissions credits remains low. As discussed earlier, the flexibility of cap-and-trade enables companies that can cheaply or easily reduce emissions to sell allowances to other companies for which such reductions are more expensive or difficult. It is possible that if the price of credits remain low, it will be more cost-effective to emit GHGs, negating the point of such schemes in the first place.

The Western Climate Initiative (WCI) is set to launch in 2012 and wants to cut emissions 15% below 2005 levels by 2020. It covers seven

US states – California, Arizona, New Mexico, Utah, Oregon, Washington and Montana – as well as four Canadian provinces – Ontario, Quebec, British Columbia and Manitoba. That makes it one of the most significant economic blocs, as the agreement covers nearly 75% of the Canadian economy and over 20% of the US economy.

Under the scheme, emissions from power generation, industry and eventually transport, are to be capped on 1 January 2012. It proposes to auction a minimum of 10% of allowances in 2012 and 25% by 2020. The states and provinces can decide individually whether to auction beyond the minimum levels. It also set a limit of 49% for the use of offset credits to reach compliance targets, which may well force significant changes to operations outside the pressure of the carbon price. Each state and province will be able to tighten programme requirements through legislation or administrative action over the next few years. It's not yet clear how many permits will be given to installations, as states can gift up to 90%, or which offsets will be fungible with the rest of the market.

The Midwestern Regional Greenhouse Gas Reduction Accord (MRGGRA) is set to launch mid-2010 and plans to cut emissions by Intergovernmental Panel on Climate Change (IPCC) recommended levels of 60–80%. Its members include six midwest states – Minnesota, Michigan, Iowa, Kansas, Illinois and Wisconsin – as well as one Canadian province, Manitoba. Indiana, Ohio and South Dakota will join the accord as observers. How the emissions will be broken down by state and industry is yet to be agreed.

The number of markets developing in the US alone underscore one of the problems that is hampering development of the carbon market globally – and that is the plethora of programmes and standards needed for effective reporting and accountability that such regional schemes require. While some companies only operate in one state, many successful businesses operate in more than one country. The larger the number of different standards that need to be addressed, the more expensive it will become for companies to comply with such schemes.

The development of regional trading schemes in North America, and indeed the promised launch of a federal scheme, are a vital part of the development of a global carbon price. There is a real need to think long and hard about how we're going to turn a patchwork of programmes into one unified system. It's possible that the US experience of integrating state and federal schemes could provide a blueprint for the eventual development of an international trading network. Many experts believe that existing state plans will be subsumed into a federal programme, but perhaps they will continue operating with credits from each programme fungible with the other.

China

China has a critical role to play in the battle against climate change. It is arguably the world's largest GHG emitter, one of the world's largest economies and has one of the highest growth rates.

China's remarkable economic development has taken two forms: firstly, rapid growth at a rate that has far outstripped the world's other leading economic powers: and secondly, a transformation from an entirely state-run economy to pseudo-free market capitalism, where the state controls only around one-third of the economy directly.

As GDP growth has been driven primarily by output rises in heavy industry and manufacturing, satisfying the increased need for primary energy has meant a massive increase in domestic coal production, particularly for electricity generation. China is dependent on coal for about two-thirds of its energy use, which has caused it to rise quickly in recent years as a major emitter of GHGs. Despite this, on a per capita basis, China's emissions remain far below those of the US.

The environmental challenges caused by China's dependence on coal are huge. According to research conducted by a team at Tsinghua University led by environmental scientist Honghong Yi, China emits more sulphur dioxide than any other nation in the world, predominantly due to its wide-spread use of coal. Acid rain falls on more than a third of the country, rain that is far more acidic than that of North America or northern Europe.

This has not only led to the loss of agricultural land, already under pressure due to the country's rapid urbanization but exacerbated the health implications of industrial pollution. A 2006 survey by the World Bank found that 16 out of the 20 most polluted cities in the world are found in China and it is believed that around a third of China's urban population is exposed to severely polluted air. This contributes not only to respiratory illness, but has been strongly linked to higher rates of cardiovascular disease.

On the basis that these economic operations continue, the World Bank has estimated that, by 2020, China will be paying $390 billion to treat diseases indirectly caused by burning coal and that this would account for an astounding 13% of its predicted GDP of over $21.1 trillion at that time. This figure does not even take into account the potential economic impact of climate change.

The main driver of growth in China's energy demand is the Government's stated aim to move its population out of poverty. The long-term goal is to raise per capita GDP to $3,000 by 2020 (from $1,000 in the year 2000), which requires growth of at least 7.2% annually. Between 1980 and 2000, China's GDP quadrupled, pulling roughly 50 million of its 1.3 billion people out of poverty, but energy consumption also doubled.

Since 1996, well over 300 million Chinese have moved from the country to the cities and this change from a rural to an urban economy is also driving the increased need for energy. China is struggling with an inadequate supply of electricity – current estimates suggest that the country has a shortfall of 80.0TWh a year, out of nationwide demand of 2,400TWh. The real figure for energy demand is unknown – there are chronic power shortages across the country due to lack of fuel.

China is focusing on increasing its clean energy technologies for exactly the same reasons as economies in the developed world. The country has energy security concerns, existing energy supplies are failing to keep up with demand (with rolling blackouts frequent) and there is a negative health and economic impact from the pollution generated in China's dependence on coal.

China's actions

In 2005, China announced its Medium- and Long-Term Energy Conservation Plan, the first phase of which – the 11th Five-Year Plan – runs from 2006 to 2010. Drawn up by the National Development and Reform Commission (NDRC), one goal is for China's total energy consumption to be held below 3.0 billion tonnes of coal equivalent by 2020 by enhancing energy efficiency.

Sector-specific consumption comparisons show that China's efficiency is far below developed country averages: thermal power is 22% less efficient; steel making is 21% less efficient; cement production 45% less efficient; and use in buildings poorer by a factor of two to three. A 10% efficiency improvement across the board could result in energy demand reduction of 100GW by 2020.

The Chinese government seems to have accepted that they must save energy, increase efficiency, develop renewable energies and cut GHGs and they already have a number of policies in place addressing the fight against global warming. China has outlined plans to achieve targets for the amount of renewable energy it produces, including an increased use of wind, nuclear and hydro power, as well as more efficient coal combustion.

China has said it hopes to increase its use of renewable energy from 7% to 10% of energy by 2010, which should be around 350GW of generation, at a cost of around US$180 billion. Its 2006 Renewable Energy Law included both quotas and feed-in tariffs to encourage development and the intention to achieve 15–20% renewables by 2020 (also similar to EU targets). At the same time, China has committed to reducing its energy intensity, or energy used per unit of GDP output, by 20% by 2010. It hopes to achieve a further 20–30% improvement in energy intensity by 2020. This is something that will affect all industry.

March 2009 saw Chinese prime minister Wen Jiabao announce an increase in spending of 146.1 billion Yuan ($21.4 billion) for climate change. This investment is considered part of the 4 trillion Yuan ($586 billion) fiscal stimulus package the government announced in November 2008. The details of the Chinese stimulus package remain to be seen but it has been suggested that 30% of those funds will go to climate change. Some will be invested in wind, solar, nuclear, and clean coal technologies but the majority – 121 billion Yuan ($17.7 billion) – will be spent on increasing farm yields by 20%, as well as raising rural incomes, both measures intended to mitigate a food crisis potentially resulting from climate change.

According to the Climate Group's report *China's Clean Revolution*, China's transition to a low carbon economy is already well underway. Despite its coal-dependent economy, says the report, the Chinese government and businesses have embarked on a clean revolution that has already made it a world leader in the manufacturing of solar photovoltaic technology (solar PV). It is likely to become the world's leading supplier of wind turbines and has the potential to compete in other markets, including solar water heaters and energy efficient appliances.

The big question is whether or not this is going to translate to actionable agreements on the international stage. According to a report from the Chinese Academy of Sciences, on its current trajectory China's emissions could double by 2030. If that's the case, it will need significant financial and technological support if it is to have any chance of altering its current economic path.

At the November 2008 Asia-Europe Meeting (ASEM), China called upon the developed world to commit at least 0.7–1.0% of GDP to the battle against climate change. The Chinese position is that the mitigation funds proposed in 2008, $6 billion in two Climate Investment Funds (CIF), amount to virtually nothing in terms of what is actually required. The issue of who is to fund mitigation and adaptation in the developing world is a contentious one, but the sands seem to be shifting and many believe that China is increasingly prepared to negotiate on targets.

China's role is also crucial in understanding one of the most difficult aspects of tracking and managing carbon emissions. It is clear that China's emissions have grown rapidly, will continue to grow and that its energy industry is dominated by fossil fuels. Continuing emissions growth in China is an argument used for inaction by many – after all, why bother to act if any cut in emissions is negated by growth in China? Yet, what exactly is driving the increase in emissions?

'Why bother to act if any cut in emissions is negated by growth in China?'

Economic growth is an easy answer, but when you start to examine the realities more closely, the facts become more uncomfortable. As we discussed in Chapter 3, according to the latest research, one-third of China's territorial emissions come from producing exports. Half of China's emissions growth between 2002 and 2005 was due to export production; another third of that growth came from capital investments, with a significant share of this in export industries. Indeed, only 15% of the emissions growth between 2002 and 2005 was due to household and government consumption.

One of the reasons that many developed economies point to China and demand shared action on growing carbon emissions before negotiating the next round of international climate change agreement, is fear. Many industries are concerned that China's less stringent environmental legislation and lack of binding emissions targets, gives the country's corporations a competitive advantage. Imposing national border taxes, however, (or some other such balancing financial approach) is only likely to kick off a trade war.

There is an alternative, which is stimulating investment in China in goods which have a carbon benefit, such as low carbon technologies, rather than cheap, disposable goods.

There is no reason why China cannot become a world leader in developing and using clean technologies. Using the latest technologies now, will greatly reduce mitigation costs later, as will investing in urban planning and infrastructure that moves China away from a fossil-fuelled future and it seems as if the government agrees. Steve Howard, chief executive of the Climate Group, says 'The reality is that China's government is beginning to unleash a low carbon dragon which will power its future growth, development and energy security objectives'. According to the report, investment in renewable energy in China is almost level with Germany, as a percentage of GDP, and the country already leads the world in installed renewable energy capacity at 152GW.

If China is to be encouraged to enter into an international climate change agreement, then surely it would be sensible to use policies that would support Chinese development, such as encouraging the manufacture of low carbon technologies cheaply. The implementation of strict global energy efficiency standards for appliances, wind turbines, solar panels and electric cars that could all be created in China's manufacturing powerhouse. Proceeds from such production could be used by the government to implement clean technologies domestically for the benefit of the Chinese people.

This is one way in which China could be incentivized to join a post-Kyoto climate change agreement. However, given the strides it has already made so far, such an agreement may not be necessary for the transition of China's economy to a low carbon framework.

Cities

Cities are particularly vulnerable to the impact of climate change. Many are based in coastal areas which, if sea levels rise, could cause a significant problem. Out of the world's largest cities, 20 out of 30, such as London, New York and Shanghai are built in low coastal regions.

They cover less than 1% of the Earth's surface but are disproportionately important in tackling the causes of climate change. About 50% of the world's population live in cities, a figure that is expected to reach about 60% by 2030. Because of their roles as hubs of commerce, as well as home to millions, through buildings, power consumption, waste, manufacturing and transportation, cities use around 75% of the world's energy.

If we are determined to move from a high-energy use, wasteful economic model to one that conserves energy and minimizes waste, then we have to implement new means of managing power generation, transportation and commerce. Cities can effect change in a number of different ways. They can impose standards and regulation on residential and commercial buildings; they regulate and sometimes manage transportation systems; they are responsible for the collection and processing of waste and, most importantly, they have significant procurement power.

Cities can negotiate great deals with the private sector, provide incentives to corporations and individuals to take action on climate change and they can stimulate and support community action. The concentration of people and resources in one place gives cities significant opportunities to introduce savings, using energy, materials and land in an efficient manner.

The best way to make a city sustainable and low carbon is to design it that way from scratch. The best known example of an attempt to do just that is Masdar City, part of Dubai's multi-billion clean technology development initiative.

Of course, for most cities it's far too late to plan for the perfect urban environment. What they need to do is plan around existing infrastructure and see how best they can minimize their impact on the environment. This is gradually beginning to be recognized. Over 350 European cities signed an agreement in early 2009 with the goal of cutting emissions by 20% by 2020. The agreement will affect 80 million people and is expected to result in €8 billion in energy cost savings.

The power of procurement

One of the key opportunities available to cities is the use of green-procurement standards, stimulating the market for low carbon technologies, goods and services, which is intended to bring down the price of goods as the supply increases.

Masdar City (UAE)

In 2007 the United Arab Emirates announced the first carbon-neutral city development, Masdar City, under its $22-billion Masdar Initiative. The goal is to build a zero emissions city by 2016, based on the planning principals of a walled city, using existing technologies to achieve a carbon-neutral and zero-waste community. The city will be connected by light rail to Abu Dhabi, which will provide transportation for around 40,000 commuters per day into the city, where there will be 1,500 businesses (with an emphasis on sustainable and alternative technology companies) and homes for 50,000 residents. The first phase of the eight-year project should be ready by the end of 2009.

The idea is that efficient urban design will eliminate 70% of the energy demand of an average non-sustainable community of similar size. It will be oriented northeast to southwest in order to ensure the optimum balance of sunlight and shade, power for construction will be generated through a large PV power plant, which will then be used to expand the city itself. Over time, the city's power will be generated through PV panels, cooling will be provided through CSP and clean water provided through a solar-powered desalination plant. Gardens, plants and trees within the city, as well as crops grown outside the city, will be irrigated with grey water and treated waste water produced by the city's water treatment plant. Grey water, domestic waste water from baths, sinks, washing machines and so on, can be used for irrigation, flushing toilets and other purposes, in order to cut water consumption.

Masdar City's design and development has been based on the One Planet Living principles, which are as follows:

- Zero carbon: 100% of energy supplied by renewable energy – Photovoltaics, concentrated solar power, wind, waste to energy and other technologies.
- Zero waste: 99% diversion of waste from landfill (includes waste reduction measures, re-use of waste wherever possible, recycling, composting, waste to energy)
- Sustainable transport: Zero carbon emissions from transport within the city; cutting air and noise pollution and congestion.

One of the first significant actions undertaken by C40 Cities, a group of cities that have committed to join forces and share experience and best practice, was the Energy Efficiency Buildings Retrofit Programme, developed by the Clinton Climate Initiative in 2007. The programme brought together 4 of the world's largest energy service companies (ESCOs), 5 of the world's largest banks and 15 of the world's largest cities in a landmark programme designed to reduce energy consumption in existing buildings.

An initial group of 15 of the world's largest cities that agreed to participate and offer their municipal buildings for the first round of energy retrofits are: Bangkok, Berlin, Chicago, Houston, Johannesburg, Karachi, London, Melbourne, Mexico City, New York, Rome, São Paulo, Seoul, Tokyo and Toronto. The international companies Honeywell, Johnson Controls, Inc, Siemens and Trane agreed to conduct energy audits, perform building retrofits and guarantee the energy savings of the retrofit projects. Meanwhile, ABN AMRO, Citi, Deutsche Bank, JPMorgan Chase and UBS committed to arrange $1 billion each to finance cities and private building owners to undertake these retrofits at no net cost, doubling the global market for energy retrofit in buildings. The financing is repaid from guaranteed future energy savings.

The programme is intended to provide both cities and their private building owners with access to the necessary funds to retrofit existing buildings with more energy efficient products. This will provide additional controls in buildings, without the need for complete refurbishment or rebuild and is expected to lead to energy savings between 20% to 50%. Not only does the programme benefit the cities themselves, but it stimulates growth in one of the most important carbon cutting measures available today – energy efficiency.

Carbon footprinting for cities

Obviously, on a city-wide scale, administrators need to understand the carbon footprint of their activities. In 2008, the Carbon Disclosure Project (CDP) launched one of the first independent programmes to help 30 cities in the US assess their GHG emissions and other climate change-related data in a standardized and comparable way.

Under the new CDP Cities Programme, at least 30 urban centres, including New York, Las Vegas, Denver, West Palm Beach, St Paul and New Orleans, will use the CDP system to assess their carbon footprint and better understand the risks and opportunities posed by climate change. CDP is partnering on the project with ICLEI – Local Governments for Sustainability USA – an international association of local governments that is driving emissions reductions and sustainable development with more than 450 members in the United States.

The Brookings Institute in Washington, DC released an analysis of the per capita footprint of the largest 100 metropolitan areas in the US, based on residential energy and transportation. The two major sources of carbon-emissions data used in the analysis were the 2000 and 2005 US Department of Transportation statistics on highway usage, as well as information obtained from Platts, a subsidiary of McGraw-Hill, on residential energy usage, but the rankings did not include traffic on local roads or the emissions of commercial, industrial, or government users.

The data is expected to help analyse emissions in the transportation systems, home heating and cooling and power generation, but it is hoped that the CDP/ICLEI analysis will be more comprehensive. Each city will assemble comparable carbon emission data within their jurisdiction's operations – for instance, fire department, ambulance and police services, municipal buildings, waste transport and other services the cities provide or activities over which they exercise budgetary control. They will also follow CDP systems to assess and disclose climate change-related risks and opportunities relating to the whole city. CDP is planning to release its initial findings in 2009.

Climate action plans

Under the C40 Cities banner there are a number of cities that have already set out climate change plans, one of which is London. The

UK is the world's eighth largest emitter of CO_2 and London is responsible for 8% of these emissions, producing 44 million tonnes of CO_2 each year. London's emissions are projected to increase by 15% to 51 million tonnes by 2025 (excluding aviation, which actually accounts for 34% of London's carbon footprint).

The London Climate Action Plan was launched in 2007 and was intended to show how change could be achieved without necessarily sacrificing standards of living and even improving the environment in which we all live, through cleaner air, increased green spaces and supportive communities. With a focus on efficiency, many of the planned measures are also intended to deliver net financial benefits in the medium term.

Stabilising global carbon emissions at 450ppm on a contraction basis means that London has to limit the total amount of CO_2 it produces by 2025 to about 600 million tonnes. Meeting this CO_2 budget will require ongoing reductions of 4% per annum. This implies a target of stabilising London and the UK's emissions at 60% below 1990 levels by 2025.

London intends to meet those targets in a number of different ways. Emissions from existing homes (40% of the total) are to be cut by 7.7 million tonnes by 2025, mostly through behavioural change such as using energy efficient lightbulbs and reducing the carbon intensity of power supplied to homes. It has been suggested that this could cut energy bills by £300 per year per average household by 2025. In 2007 former London Mayor Ken Livingstone also launched the Green Homes Programme, offering subsidies for insulation, advice and training. Another programme for companies is also in the planning, once again with an expectation of cutting 7 million tonnes of CO_2 by 2025. This will focus on upgrading commercial buildings, working with tenants to reduce emissions and lobbying for further government initiatives. Another 1 million tonnes of CO_2 per year in 2025 is to be saved through the enforcement of current regulations and improved standards for new build.

The plan also calls for emphasis on energy efficiency through planning, greater focus at a borough level and a transition of its power supply from the National Grid towards a local, lower carbon-energy supply, including CHP, energy from waste and on-site renewables like solar. The goal is to move a quarter of London's energy supply off the grid and on to local, decentralized systems by 2025, with the majority of London's energy being supplied in this way by 2050.

London is unusual compared with many large cities around the world in that its transport emissions (excluding aviation) are relatively small – about 22% of the total. There are high levels of public transport use and policies, such as the congestion charge, have been used to manage traffic. If fully implemented, the traffic measures in the plan, including promotion of alternatives to car travel, increases in vehicle efficiency, the promotion of low carbon fuels and carbon pricing for transport, would deliver carbon savings of 4.3m tonnes by 2025.

Many cities have instigated programmes which can be used as an example by others on how to tackle climate change. It is useful to consider a few case studies which involve the use of differing approaches and concentration on various sectors. What they have in common is a structured approach to addressing specific problems within city frameworks, in ways that result in lower emissions.

São Paulo and waste

As part of a wider GHG reduction plan, the City of São Paulo has developed landfill gas to energy plants on the two largest landfill sites in the world – Bandeirantes and São Joao.

The Bandeirantes plant has the potential to generate 170,000 MWh, enough power to supply 400,000 people. The landfill receives about half the daily waste produced by the City of São Paulo, which is about 7,000 tonnes of waste every day.

The São Joao landfill reached capacity at the end of 2007, having received almost 28 million tonnes of waste over its lifetime. Since

1992, it had received a daily average of 6,000 tonnes of rubbish and produced 1,800 cubic metres per day of liquid that was previously sent for treatment at the state water treatment plant.

The projects are environmentally sustainable as waste no longer contaminates underground water, but only the ground. Once the project is completed the landfill itself will be remediated and contamination levels reduced to zero. The use of the methane to generate power could cut the emission of 8 million tons of CO_2e over the next 15 years – methane has a global warming potential between 23 and 26 times that of CO_2.

According to the São Paulo government, the plants will generate sufficient power for 7% of the city and will prevent the emission of a total of 11 million tonnes of CO_2e by 2012. At the same time, an area of environmental protection of 270,000 sq m surrounding the landfill was created, which is being used by the local community for environmental education programmes.

The city benefits from this project in a number of ways, including the use of landfill sites for power generation and the generation of extra financing for the city from the sale of its carbon credits.

San Francisco and buildings

The city's Climate Action Plan was adopted in 2002 and in doing so, San Francisco joined over 500 cities participating in the Cities for Climate Protection Campaign of the International Council for Local Environmental Initiatives (ICLEI). Initial research found that energy use in buildings and facilities is responsible for approximately 50% of San Francisco's GHG emissions. In 1990, San Francisco's energy use resulted in a total of approximately 4.5 million tonnes of CO_2 emissions being released into the atmosphere, making green building a critical component in the fight against climate change.

In 2008, a new green building ordinance was implemented in San Francisco. This specifically requires newly constructed commercial buildings over 5,000 sq ft, residential buildings over

75 ft in height and renovations on buildings over 25,000 sq ft to be subject to an unprecedented level of Leadership in Energy and Environmental Design (LEED) standards and green building certification, making San Francisco the city with the most stringent green building requirements in the US.

Some of the significant cumulative benefits this ordinance is expected to achieve through 2012 are: reducing CO_2 emissions by 60,000 tonnes, saving 220,000 MWh of power, saving 100 million gallons of drinking water, reducing waste and storm water by 90 million gallons, reducing construction and demolition waste by 700 million pounds in weight, increasing the valuations of recycled materials by $200 million, reducing automobile trips by 540,000 and increasing green power generation by 37,000 MWh.

While the direct benefits of the programme are clearly improvements in San Francisco's emissions profile, it will also benefit from a reduction of consumption in a wide range of other resources.

Seattle doubles energy conservation

In 2008, Seattle Mayor Greg Nickels launched an energy savings programme, Initiative 937, to double Seattle utility Seattle City Light's energy conservation programme over the next five years. The $185 million investment in dozens of efficiency programmes is expected to save customers more than $310 million in bills over five years and create more than 1,000 green jobs.

Energy demands on Seattle City Light are expected to increase more than 1% annually. Initiative 937 demands that electric utilities with more than 25,000 customers ensure that 15% of their power comes from new, renewable energy sources by 2020. More than wind, land-fill gas, or geothermal, Seattle City Light will meet its needs through conservation, which is also counted under Initiative 937.

According to Seattle City Council, since conservation efforts began 30 years ago, customers have saved more than half a billion dollars, with an estimated $63 million in projected savings for 2008 alone. The Five-Year Conservation Action provides for 1 million tonnes of avoided CO_2 emissions and it's estimated that the new residential and commercial conservation programmes could result in more than 1,000 local green jobs such as auditors and heating and cooling technicians.

The city has implemented programmes that have shown themselves to cut costs in energy bills and act as frameworks to encourage further investments in technologies and jobs in the energy efficiency sector.

Rizhao and solar heating

Since 2001, Rizhao, a city of 3 million people in northern China, has been promoting the use of solar energy to provide energy, heating and lighting. An incredible 99% of Rizhao's households use solar water heaters, while almost all traffic signals, street lights and park illuminations are powered by photovoltaic solar cells. In total, the city has over a half-million sq metres of solar water heating panels, the equivalent of about 0.5MW of electric water heaters.

Rizhao has a lower per capita income than some neighbouring cities, so Mayor Li Zhaoqian focused on increasing the efficiency and lowering the cost of solar units. Rather than provide subsidies to solar panel users, the government funded industry research and development. Solar water heaters cost the same as electric alternatives, roughly $190 (4–5% of the annual average income of a household in town, 8–10% of a rural household's income) and save users money in the long run.

The city then mandated that all new buildings incorporate solar panels and oversaw the construction process to ensure proper installations. Government buildings and the homes of city leaders were the first to have the panels installed. Some government bodies and businesses provided free installation for employees, although the users pay for repairs and replacement.

Consequently, 99% of households in the central district obtained solar water heaters. Using a solar water heater for 15 years costs about $1,934 (15,000 Yuan) less than running a conventional electric heater, which equates to saving $120 (Yuan) per year per household.

The implementation of solar heaters has had a direct impact on energy consumption. At the same time, it has helped to stimulate the local economy. In 2007, the city received one of the first World Clean Energy Awards for the project 'Popularization of Clean Energy in Rizhao'.

Cities, by virtue of their responsibility in setting standards for efficiency or development, managing transportation and waste management networks, and in deciding to offset their carbon, can play a significant leadership role in the transformation to a low carbon economy. Outside London, none of the projects mentioned above were in countries with a binding emissions target at the time the project was undertaken. What these projects reflect is the wider implications for economics, clean air and comfort that addressing emissions and streamlining energy use and transportation can have. It is also worth noting that the financing associated with the generation of carbon credits can make a significant difference to the projects undertaken.

Corporations

Governments may implement legislation to drive change, but it is corporate action that enables the development and deployment of low carbon technologies. Products and services are the front line in the creation of low carbon technologies. There are many reasons why companies are beginning to take a more active role in addressing environmental issues.

Industry lobbyists have expressed concern about the detrimental long-term impact of carbon constraints on shareholder value and profitability, especially with regard to expected increases in energy prices and the potential impact of regional carbon trading on global competitiveness. They're worried about the increasing burden of complying with new environmental regulations and about how their reputations could suffer. Yet there is also a potentially enormous economic opportunity to not only create new products, but to create new ways of manufacturing and distributing those products.

Public corporations are responsible to their shareholders for making as much profit as possible. While the last 10 years has seen an increasing number of companies take a stand on corporate and social responsibility, as a general rule, the profit motive is key. Shareholders have traditionally been focused on returns but the implications of climate change are beginning to have an effect, both on corporate interest in cutting their own resource consumption and in shareholders taking a wider picture with regard to corporate activity.

Corporate social responsibility (CSR) has played an increasing role in corporate activities over the last decade or so. The idea is that corporations should take into consideration how their operations impact on quality of life, health, learning, inclusion and development. The central notion is that sustainabilty should lie at the heart of corporate decision making because it's one of the most important forms of long-term risk management.

There are business risks associated with climate change, ranging from problems in sourcing commodities, impact of fuel price rises, to inability to source sufficient clean water (critical for many producers of food and drink). All of these will affect the ability of a company to do business. The impact of climate change and carbon regulation can also add another level of potential financial risk to a business and it's critical that these issues are understood by investors.

There is enormous economic potential in new markets being driven by the need for new low carbon products and services, as well as new legislation and standards. The renewable energy industry has seen billions of dollars worth of investment in the last few years and that is set to increase. Every industry affected by the need to decarbonize the economy will see investment and growth, as well as employment opportunities. The potential ranges from energy generation all the way down to the level of individual appliances.

According to DEFRA, about 80% of the environmental impact of a product is defined at the design stage, which means that it is in the hands of industry to create sustainable products.

The EU Framework Directive on Energy-using Products (EuP Directive), which was launched in 2005 legislated on the design of energy-using products, requiring an assessment of the GHG emissions associated with product life-cycle. A requirement for appliances to be increasingly energy efficient drives the market for the appliance and for the components that allow its efficiency to be increased. There are plans to introduce such standards into the information, communications and technology (ICT) market and standards under review at the time of writing include a requirement to reduce the energy demand of lighting equipment by 15%, while TV boxes should require nearly three times less power by 2014.

The law requires that all white goods, such as fridges, freezers and washing machines are labelled, assessing the product's energy efficiency and giving an estimate of electricity and water consumption on standard settings, performance, noise levels and whether the product has gained an ecolabel. The label identifies the most energy efficient models on a scale of A-G, A being the most efficient.

In 2007, General Electric (GE) chief executive Jeff Immelt announced that the company was commiting to the development and deployment of new solutions to environmental challenges, with products ranging from efficient gas turbines, wind turbines, batteries for transportation and the grid, energy efficient lighting, solar panels and water purification technologies. Originally GE said it expected to sell $20 billion in Ecomagination products and services by 2010. In 2008, GE said it expected annual revenue for such products to be up to $17 billion in that year alone.

The capital markets have been structured to reward success and many believe that if the right demands are made, the business sector has the entrepreneurship and the capital resources to marshall the investment needed for an effective transition to a low carbon economy.

As things stand one can be reasonably sure that one of the following is the reason that a company becomes environmentally active.

- Legislation
- Changing economic environment
- Shareholder pressure
- Saving money
- Making money
- Brand, reputation and customer differentiation

We're going to look at the actions of a number of companies in different sectors and countries to see how they're taking action on lowering their carbon emissions.

Wal-Mart

The retail giant Wal-Mart has about 20% of the US grocery and consumables market and owns Walmex in Mexico, Asda in the UK, as well as operations in a number of other countries. In 2007, it started working with the Carbon Disclosure Project (CDP), to collect emissions data from all its suppliers and within its operational management within the US. The idea was that once the company had measured that data, it could do something about managing emissions. As the largest private electricity user in the US, Wal-Mart's first action to tackle climate change has been through energy efficiency.

Once it had clarified the data, it then worked with The Climate Group to cut its emissions, which resulted in significant savings. Calculations showed that in 2006, the company's carbon footprint was 19,136,882 metric tonnes of CO_2 equivalent.

In response to the findings, the company set a number of targets:

- Cut energy use at its 7,000+ stores worldwide by 30%.
- Cut GHG emissions by 20% at existing stores in 7 years.
- Sell 100 million compact fluorescent light bulbs per annum by 2008.

- Increase fuel efficiency of trucking fleet by 25% over 3 years and double efficiency within 10 years.
- Reduce packaging by 5% by 2013.
- Reduce solid waste by 25% from US stores within 3 years
- Transform Wal-Mart into a company that runs on 100% renewable energy and produces zero waste.

While reaching those targets may take some time, the company has already taken significant strides. It invests around $500 million per year in energy-saving technologies. In one example, the company spent about $30 million to develop a refrigerator LED lighting system with General Electric and Royal Philips Electronics, which cut 50% of the cost of the lighting. Another key area is the cost and operation of heating and air conditioning systems (HVAC). Specifically, it is looking at the integration of new HVAC with efficient refrigeration systems, so that 100% of the heat rejected by the refrigeration system is reclaimed into the HVAC.

Wal-Mart has said that it hopes to save up to $3.4 billion by reducing packaging by 5%, as part of an overall supply chain saving of $10.89 billion. Cutting packaging, aside from cutting waste and requiring lower inputs, also has an effect on other operations. In one example, by reducing the packaging on fewer than 300 toys Wal-Mart saved $3.5 million in transportation costs alone.

The company has started measuring energy use and emissions within its supply chain. Wal-Mart has around 60,000 suppliers so it's a big job, but reading the electricity and gas metres at 60,000 factories is going to be a lot easier than measuring the carbon footprint of each product. It should be said that the company's position is very clear in emphasising the role that energy efficiency and conservation could have in reducing supplier costs and therefore prices in its stores. While the end result may be an overall improvement in the company's impact on the environment, the activities have predominantly been driven by a cost-benefit analysis.

Tetra Pak

In 2007, Tetra Pak produced 137 billion drinks cartons equating roughly to a demand for 92.5 million trees to be used for paper and pulp. According to the US Environmental Protection Agency (EPA), packaging accounts for roughly one-third of commercial and municipal waste. In 2007, the UK sent roughly 10 million tonnes of packaging waste to landfill; the previous year, the US equivalent was around 80 million tonnes. Increasing demand for packaging globally is putting pressure on supplies, mostly timber: this has led to an increase in illegal logging, a matter of enormous climate concern.

In 2005 Tetra Pak signed up to the WWF's Climate Saver Programme, in which members have committed to reducing their CO_2 emissions by 10 million tonnes per year by 2010. Tetra Pak pledged to cut its emissions by 10% by this date. The company agreed to invest €6 million in energy efficiency and €3 million extra in the use of green energy. It has facilities in the US and Denmark that are fully powered by renewables and the company offsets its fossil-fuel emissions with credits using the WWF's Gold Standard.

Part of its agreement was to join the WWF Global Forest and Trade Network. This has resulted in the majority of the wood fibers and paperboard used in its cartons being certified as sourced from plantations with forestry management standards under the rules of the Forest Stewardship Council and WWF Global Forest and Trade Network. The FSC, which requires forests to protect biodiversity and indigenous livelihoods, adhere to local and international law and practise forest renewal programme, certifies 25%. The company is streamlining manufacturing efficiency, through the introduction of new filling machines and is working on a number of local schemes to improve recycling. Another approach it took was in the streamlining of transportation of its products, cutting emissions as a result.

While the cost benefit is clear from the company's actions in terms of decreased fossil-fuel costs, transportation costs and waste, it is also apparent that the company's activities were driven by a desire to reposition itself as a sustainability leader.

Google

While one might not think that a company which operates a purely online business would have to worry much about its contribution to CO_2 emissions, the power consumption required to keep its servers online and therefore its service up and running, is enormous. Working with the Climate Group, the company set itself a goal of being carbon neutral by the end of 2007. While it is making active strides in increasing efficiency and use of renewables, to date it has declined to disclose the emissions that it is actually trying to neutralize.

The company has worked both to cut its own operational emissions and to invest in carbon cutting programmes through Google.org. To help reduce global emissions related to computer use, Google joined the Climate Savers Computer Initiative. By 2010, the Initiative's goal is to reduce global computer CO_2 emissions by 54 million tonnes per year.

To increase onsite generation of renewables, Google installed solar PV panels on Googleplex rooftops and on two of its carports. Google expects to save more than $393,000 annually in energy costs – or close to $15 million over the 30-year lifespan of its solar system. EI Solutions, the company that designed and installed Google's PV system, estimates the system will pay for itself in approximately 7.5 years.

Google launched a free car-sharing programme for its employees, offered a US$5,000 rebate to employees who purchase a vehicle that runs at over 45 mp/g and is working on improving the energy efficiency of its data centres. It is looking to do this through both design improvements and the adoption of power-saving technologies, such as evaporative cooling.

In one example, Google.org launched a programme to reduce emissions from transport through the Recharge It Program – an initiative that aims to reduce CO_2 emissions, cut oil use and stabilize the electrical grid by accelerating the adoption of plug-in hybrid electric vehicles and vehicle-to-grid (V2G) technology. As part of this initiative, Google.org awarded $1 million in grants and announced plans for a $10 million request for proposals (RFP) to fund development, adoption and commercialization of plug-ins, fully electric cars and related V2G technology.

For Google, it's clear that the actions of the company are being driven by philanthropic, cost and environmental factors. One thing to be cautious about, however, is the fact that we don't know the company's total emissions and we don't know what emissions are being included in the company's calculations – it's difficult to get a clear picture of what the company is achieving.

The Savoy Hotel

It's not just multinational firms and massive public-sector bodies that can benefit from the introduction of environmental and efficiency standards. The Savoy Hotel in London, one of the world's most iconic hotels, has implemented a carbon reduction strategy that has resulted in a 40% cut in its energy consumption.

Its carbon-reduction strategy was part of the hotel's 2007–2008 £100 million restoration project, which will see The Savoy reduce its electricity consumption – the greatest source of its carbon emissions – by approximately 50% and total energy consumption by up to 40%. This is expected to result in permanent emissions reductions of 3,000 tonnes of CO_2 annually.

Working with Evolve Energy, a specialist energy management group, the hotel has transformed its energy management. Energy savings devices in the new Savoy will include the replacement of old boilers and heat transfer equipment with ultra high efficiency hot water and heat recovery systems, low energy air conditioning, intelligent building controls, smart metering and a 24-hour monitoring and targeting of energy use at the hotel. Every room and each suite will be fitted with useful energy saving equipment, including smart guestroom controls, as will the public areas.

There are clearly opportunities for companies to save money within their supply chains and to lead the movement towards a more sustainable framework. However, one of the key things that needs to be resolved are the standards to which companies should be held. The current market, in terms of financial stability, market and consumer expectation, regulation and legislation is volatile. There is an opportunity to base operations on both responsibility to stakeholders and opportunity to grow, but the market needs a greater understanding of what is expected.

The danger in going green

Given changing market expectations, companies run the risk of damaging their reputations in the eyes of the public by failing to make alterations to their behaviour or falling out of step with their customers on the issues. While a few companies are publicly taking action on moral or ethical grounds, the profit motive is increasingly driving action in today's corporations. The biggest difficulty in transforming business practice and communicating that effectively to stakeholders, is the general confusion about what makes a company green.

The growing use of 'green' terms in marketing products and services is causing confusion. 'Greenwash' is a term that has been coined to describe the inappropriate use of green terms to sell products. Terms such as 'sustainable', 'green' and 'clean' are frequently used, yet there is not a globally accepted definition of what they mean, or where the boundaries of their use should lie. Overuse of green terminology and imagery has led to a growing consumer backlash.

The 'green' market has a two-fold problem – that many of the terms bandied about on a regular basis are confusing or ill-defined and that those terms are often used by companies to promote goods and services without a real environmental benefit.

Sustainable development has been defined as 'development which meets the needs of the present generation without compromising the ability of future generations to meet their own needs', but the terms within the definition can mean different things to different people. The need or desire to integrate economic, environmental and social considerations into every business decision is great, but what if you give different weighting to the issues under consideration?

The term 'carbon neutral' has similar problems. What do we mean by neutral and how do we achieve it? It's effectively taken to mean a combination of energy efficiency (the reduction of energy consumption), purchase of clean energy (the use of renewable energy for onsite power consumption), and the purchase of carbon credits to offset any remaining emissions. Basically, reducing, replacing and offsetting one's carbon emissions. Even if we ignore the practical issues still rampant in the voluntary offset market, until every aspect of carbon within the supply chain has been measured, how can we guarantee that we're offsetting all necessary carbon?

Talking the green talk is no simple matter, with a lack of definitions, high expectations and countless critics ready to pounce if you don't get it right. If the markets are to be as efficient as possible, we need to streamline the communication of information between companies, shareholders, suppliers and consumers. According to John Grant, in his 2007 book The Green Marketing Manifesto, what's needed is for companies and marketers to make green normal – 'it's not about making normal stuff seem green, its about making green stuff seem normal'. In other words, it's not a question of a new niche marketing opportunity, but an opportunity to reframe our expectations.

Education across industry and among consumers is going to be a key factor in eliminating such mistakes. In the US, Greenpeace has launched a campaign called 'Stop Greenwash', intended to track and highlight instances of such behaviour and outlines four qualifying criteria for identifying greenwash:

- Dirty Business
- Ad Bluster
- Political Spin
- It's the law, Stupid!

Dirty Business: This is when a company markets a particular environmental programme or product, while the corporation's product or core business is inherently polluting or unsustainable.

This can be easy to identify. For example, Oil giant Shell got itself into hot water in 2008 when it claimed that its $10 billion oil sands-project in Alberta, Canada, was going to provide a 'sustainable energy source'. The tar sands cover over 140,000 sq km of Alberta and contain nearly 173 billion barrels of oil in the form of bitumen. This is transformed into crude oil through high energy, carbon and water-intensive extraction and treatment procedures. A report from Co-Operative Investment and the NGO WWF, *Unconventional Oil: Scraping the Bottom of the Barrel*, suggested that the production of oil from tar sands can create up to eight times as many emissions as producing conventional oil, as well as consuming vast amounts of water. That doesn't sound very sustainable.

Shell argued that the need for affordable and convenient energy is a crucial part of strong economic and social development over time, one of the main tenets of sustainability. The UK's Advertising Standards Authority (ASA) didn't agree and supported WWF's complaint. However, it should be noted that it was only on the grounds that there was no data showing how Shell was managing emissions from the project, which meant that the suggestion was ambigious, not inaccurate.

Ad Bluster: This is using targeted advertising and public relations campaigns to exaggerate an environmental achievement in order to divert attention away from potential problems, or when a company spends more money advertising an environmental achievement than actually doing much to achieve it.

In 2008, Greenpeace awarded BP with the first annual 'Emerald Paintbrush' award for doing just that. During 2008 BP ran a multimillion dollar advertising campaign announcing its commitment to alternative energy sources, with slogans including 'from the earth to the sun, and everything in between' and 'the best way out of the energy fix is an energy mix.' While investment sums are relative, Greenpeace said it had obtained a presentation which reveals that the company allocated 93% ($20 billion) of its total investment fund for 2008 for the development and extraction of oil, gas and other fossil fuels. In contrast, solar power (a technology which analysts say is on the brink of a technological breakthrough) was set to receive just 1.39% ($0.3 billion).

Political Spin: This is when companies advertise or speak out about corporate green commitments while lobbying against pending or current environmental laws and regulations. As, for example, when public statements or advertisements are used to emphasize a company's focus on environmental responsibility while in the midst of legislative pressure or legal action.

Greenpeace selected General Motors (GM) as a leading example of this approach. Its 'Gas-Friendly to Gas-Free' advertising campaign in 2007 was an attempt to reposition the company as environmentally responsible and progressive. However, the company remains the leading manufacturer of high fuel-consumption cars and is reported to have been the car industry's biggest lobbyist in 2007, trying to influence members of US Congress to prevent them passing legislation to increase fuel economy standards. It's arguable that the use of green terminology to promote current internal combustion driven cars in any way is open to accusations of greenwash, as they consume, by definition, large amounts of fossil fuel.

It's the Law, Stupid!: This is when a company advertises or brands a particular product with environmental achievements that are already mandated by law. For example, if an industry or company has been forced to change a product, clean up its pollution or protect an endangered species, then uses PR campaigns to make such action look voluntary.

The American Coalition for Clean Coal Electricity (ACCCE) is identified as using this particular approach. ACCCE is a wholly owned subsidiary of the US coal industry and is supported by the

coal, rail and electricity industries: including ALCOA, American
Electric Power, CSX, Detroit Edison, Duke Energy, Peabody Energy,
Southern Company and Union Pacific Railroad. Greenpeace says its
real purpose is not to promote coal as a source for clean or green
energy, but merely to ensure that the US continues to be highly
dependent on coal for its energy needs.

Although ACCCE claims that its 'coal-based generating fleet is
70% cleaner than before,' these numbers refer only to reductions in
sulphur oxide (SOx) and nitrogen oxide (NOx) emissions, which
were demanded by legislation in the 1980s. The US coal industry
has yet to implement technology to reduce CO_2 emissions, the
main cause of global warming.

Addressing the problem of greenwash

One of the key problems with greenwash is the antagonism that it
can generate. The confusion created for stakeholders, especially
consumers, can lead to disillusionment. There are companies
making positive improvements in their energy supply and
procurement practices. They don't promote their green initiatives,
however, on the grounds that they might be attacked for not going
far enough, or for not achieving their stated goals in the time frame.

The difficulty for many corporations is understanding how to
position themselves in the green marketplace. Is it acceptable to
make moves in the right direction while having a business model
which is, by definition, anti-environmental? Or is it only acceptable
for a company to completely reinvent itself? We have to ask ourselves
whether we should demand perfection or accept moves towards a
common goal – should there be a sliding scale of expections? We
need to understand where a company is making a real effort to
achieve change and find some way of rewarding that action.

Banking and dirty coal

The involvement of major financial corporations in the financing of
fossil-fuel projects is of increasing concern. A 2008 report from

Friends of the Earth, *Cashing in on Coal* refers to an unpublished analysis undertaken by the Rainforest Action Network, showing that financed fossil-fuel extraction consistently represents more than 99% of the total carbon footprint of large commercial banks, with those emissions covered by the CDP representing less than 1% (referring to Scope I and II emissions).

This is particularly an issue in the UK, since the government backs the development of new coal-fired power stations to be ready for effective carbon capture and storage (CCS) as and when the technology is developed. Much has been made of E.On's planned development at Kingsnorth, but it is only the first of seven new coal-fired power stations for the UK. It has been suggested that the 50 million tonnes of CO_2 that these seven new plants would emit will wipe out any chance the UK has of cutting emissions by more than 80% by 2050 – the level that a majority of scientists believe is necessary to avoid disastrous climate change.

Barclays has been a prime mover amongst the banks in addressing climate change issues in both its management and its business practices. In 2007, Barclays signed a deal with EDF to provide half of its electricity from renewable sources (up from 3%), signifying its commitment to solutions to climate change. The bank opened one of the first carbon trading desks, was the first to take physical delivery of EU allowances and the first bank to take delivery of Certified Emissions Reductions (CERs).

It launched a Carbon Offset corporate charge card, as well as a consumer product, Barclaycard Breathe (a credit card donating 50% of profits to environmental projects) and it is rolling out a carbon neutral debit card to all its UK customers. Barclays Capital, the investment arm of Barclays Bank, has provided long-term finance for over 2.6GW of renewable energy generating capacity, but should this commitment offset its investment in other carbon intensive industries?

Cashing in on Coal underlined how many of the banks are highlighting their involvement in renewables while simultaneously issuing enormous and lucrative loans to coal and other energy and carbon intensive companies. The report claims that between May 2006 and April 2008 Barclays was involved in 17 loans to coal-related companies and responsible for sourcing $5.79 billion in loans to companies engaged in the extraction and/or combustion of coal. While there is no direct link made in the report between the Kingsnorth project and the bank, it does report that in 2007 Barclays, RBS and HSBC loaned E.On $70 billion.

Barclays is one of the largest banks in the world and cofounder of the Equator Principles, a voluntary code of practice agreed by 10 big banks for managing environmental and social issues in project financing. The code is intended to ensure that the banks uphold sound environmental management practices and that all projects worth over $10 million should be audited and tested against criteria including environmental damage, pollution and sustainability. They also oblige investors to ensure that negative effects on project-affected ecosystems and communities should be avoided where possible and if these effects are unavoidable, they should be reduced, mitigated and/or compensated for appropriately. Quite how large-scale coal projects with no CCS component fit this criteria is difficult to understand.

The question being asked is the extent to which Barclays' positive moves are undercut by ongoing investment practice. While it's arguable that it's economically unrealistic for major banks to move away from investment in traditional energy, it is possible – the Co-operative Bank's existing ethical investment policy prevents it from financing coal, oil or gas projects. While Barclays may have positioned itself as a leader in carbon management within its own operations, at some point soon it's going to have to address the impact of its lending and investment portfolios in exacerbating climate change – or stop positioning itself as green.

InterfaceFlor

InterfaceFLOR, on the other hand, is a company that, at first glance, shouldn't qualify as being green. The company makes modular flooring, most of which is nylon and generated from fossil-fuel based resources. Yet, the company prides itself on being one of the first companies to publicly commit to tackling climate change. Mission Zero lies at the heart of the company's strategy, a commitment made by founder Ray Andersen to achieve a zero carbon footprint and eliminate any negative impact the company may have on the environment through its operations by 2020.

Its stated aim is to become the world's most sustainable company. That covers more than energy, as the company employs full life-cycle analysis on its products, water use and product disposal. It considers the entire supply chain, from raw material acquisition, product manufacture and transport all the way through to how customers use and dispose of the products. It has also set clear boundaries regarding the emissions that it includes in its footprint, such as emissions from manufacturing, office operations and transportation of people and products.

The company has attempted to transform operations away from the traditional approach of 'take, make and waste' and towards a more natural or holistic approach. This includes the implementation of an overall equipment effectiveness scheme, a total productive maintenance scheme and even a bonus scheme which rewards employees for reducing waste and minimising downtime. Interface has also worked to reduce energy usage to an absolute minimum and has switched to renewable energy at all its manufacturing facilities in Europe. Of course, the company has benefitted as well. As a result of these initiatives, InterfaceFLOR has made significant cost and efficiency savings across the business – to date the company has saved more than \$372 million in avoided waste costs. It has also shown a net 82% reduction in CO_2 emissions since 1996.

The company provides other options as well. Its products includes Cool Carpet, where clients pay a small premium on the purchase price to offset its lifecycle emissions; as well as ReEntry, where InterfaceFLOR works with partners to take back end of life products for repurposing and recycling.

The company has also reduced its dependency on petrochemicals as a raw material and is actively researching alternatives. According to European chief executive Lindsey Parnell, one of the biggest problems lies in customer expectations. Flooring obviously has to be hard-wearing and often procurement contracts specifiy the flooring should be useable for 15 years. However, many clients refurbish more frequently than this. In order to achieve change, clients need to have a better understanding of sustainability themselves. There is a wider social issue involved, as if expectations of performance were set up in a more realistic fashion, it's possible that the replacement of materials would get around the problem.

Perhaps one of the most important actions that InterfaceFLOR has undertaken is in setting out the boundaries for its move towards carbon neutrality. Many financial services firms, in particular, have claimed to be carbon neutral, but have set the boundaries for emissions at Scope 2, meaning they ignore any indirect emission. By accepting responsibility for its entire carbon footprint and setting a goal of neutralising it all, the company has set up a powerful precedent and should be highly commended for that alone.

Innocent

Sometimes a company where operations do meet high sustainability standards can run into trouble for no other reason than poor communications. In 2008, UK-based smoothie-maker Innocent was accused of greenwash.

Innocent positions itself, very successfully, as a company that attempts to do the right thing. The company is focused on using natural and responsibly-sourced ingredients, sustainable packaging, resource efficiency and donates 10% of its profits to charity every year (mostly to the Innocent Foundation, which funds rural development projects in the countries where it sources fruit).

Innocent has worked with the Carbon Trust to work out its carbon footprint using the PAS 2050 standard. That means it included emissions from growing fruit, transportation, crushing and blending, manufacturing, packaging, bottling, distribution and refrigeration in ships. Using this method, in 2007 Innocent reduced the footprint of its 250ml smoothie by 16%. Each of its drinks displays the amount of CO_2 emissions associated with its consumption and, through its partnership with the Carbon Trust, it has committed to cutting that carbon impact over a two year period.

The company's work within its supply chain helped identify an opportunity for increasing the recycling of waste materials, which reduced waste sent to landfill by 15% within the first month, a figure that increased within six months to 54%. By environmental standards, Innocent is a market leader.

On 2 August 2008, the Daily Telegraph ran an article 'revealing' that the company's smoothies were blended in Rotterdam and then shipped to the UK, while the company's site said that they were produced in the UK. The site also said that its fruit travels by boat and rail, because that is a less carbon intensive means for transport – fewer emissions per kilo of fruit. Innocent said that it was a miscommunication as the website had simply not been updated since blending had been moved overseas.

The issue here is not whether Innocent did anything wrong – as it's certainly far more efficient to transport pulped fruit than pre-pulped – it's the standard to which Innocent is held by public opinion. This came up again in early 2009 when Innocent announced an investment by Coca–Cola and many commentators suggested surprise at the combination of such a non-corporate brand with a multi-national behemoth. What this underscores is how every aspect of a company's approach, from core business, to operational management, to type of shareholder, is open to review.

What next?

Once corporations accept a responsibility to operate in a sustainable and reasonable way in terms of resource consumption, energy use and waste management, we are already on our way to the low carbon economy that the world requires. Shareholders are putting increasing pressure on their portfolios to address the issues, and a corporation's ability to measure and manage its own emissions is likely to become a key measure of management capability in the medium term. Increasing legislation regarding pollution and emissions is going to force action on many companies, even if they are unprepared as yet.

Consumers are paying increasing attention to the actions of corporations, and demanding that they address their emissions and activities in the context of sustainable living. We are, however, at a very early stage. There are a range of criteria upon which companies can report, and it can be very difficult to compare responses from different companies.

There is a significant danger that if corporations take action but are not consistent in defining their actions, achievements and long term goals, they could be accused of greenwash. What is obvious is the need for clarity, reporting standards and better communication. This is needed for both corporations and individuals alike. Success in the low carbon era is going to require a rethink about transparency, reporting, business planning and strategy.

Corporations prefer long-term frameworks in which to develop long-term business strategies but the reality is that the science, the politics, the environment and even our economic framework are in a state of flux. Everything evolves and those corporations likely to prove most successful in the coming years will be those which are able to respond to new opportunities despite being tempered by new responsibilities.

There are clear economic benefits to be gained through beginning that transition today, an approach which can be aided by individual activity and consumer purchasing power. In which case its time that we took a look at the role of individuals in addressing the battle against climate change.

Consumers

One question that many people ask is whether, as individuals, we have any real power to influence change. If we are lucky enough to live in a democracy, we should be able to answer that question with a resounding 'yes'. Even though many of us feel disenfranchized by the political process, disillusioned by the expectations and behaviour of our economy and our society, each and every one of us is responsible is in some small way for the society in which we live. And that means that if we want to change it, we have to take action. On one level that means that until each and every consumer becomes responsible for his or her own emissions, not enough is going to change.

Influencing choices

Decisions are made for a number of reasons and at different times. People often know what they should do, such as stop smoking, drinking or eating bad food, but that doesn't mean that they find it easy to change their habits. Timing is hugely important, if we are to be asked to make the right choice at an appropriate time. One of the key things that we need to develop in order to support action on climate change is a positive choice architecture, a system by which people respond to various stimuli in well understood ways, in order to achieve a collective goal.

Countries, cities and corporations all have the ability to act as choice architects for consumers, as leaders to focus our attention in the 'right' direction. In the same way, consumers have the ability to influence companies and politicians alike, by making very clear the negative market and electoral impact of 'poor' choices.

In their 2008 book *Nudge: Improving Decisions about Health, Wealth and Happiness*, Richard H. Thaler and Cass R. Sunstein explore how little nudges can have a big impact on the decision-making process of the individual. In the authors' view, the process of decision-making is a war between the 'automatic system' (the rapid, intuitive, reptilian part) and the 'reflective system' (the slow, deliberate, self-conscious part). They define a nudge as any aspect of the choice architecture that alters people's behaviour in a predictable way without forbidding other options, or significantly changing their economic incentives. Both cap-and-trade programmes and environmental taxes are seen as 'nudges' as they don't change the economic cost of the decision to purchase a product, instead they merely alter the perceived cost, as external costs such as pollution are included in the price.

'Until each and every consumer becomes responsible for his or her own emissions, not enough is going to change.'

However, it's Thaler and Sunstein's work on information and feedback that is perhaps most interesting. We live in a world of innumerable choices, with little time and little information to make the best choice. Even if we do make the best choice, we often receive no feedback as to the impact of that choice. That's especially true on the subject of climate change.

Feedback is a critical problem in motivating change – we can't see the long-term effect of our pollution and we often don't see the short-term effect of our actions. An example of the power of effective feedback was the 1972 US National Environmental Policy Act – this required the government to compile and disclose environmentally related information before starting any project that might affect the environment. This led to significant change in corporate behaviour, as environmental groups began to target the worst offenders and such forced disclosure meant a significant reputational risk. The decision in the US by the Environmental Protection Agency (EPA) to introduce a comprehensive national reporting system for GHG emissions could well have the same effect with regard to CO2e.

Where it's more difficult is in showing what difference can be made by the small actions recommended for individuals. The fact remains, however, that every action is cumulative. For example, if every light bulb in every London home was energy efficient, it's been estimated that Londoners could save 575,000 tonnes of CO_2 and

£139 million per year; if all appliances in homes were energy-efficient, this could translate into savings of £150 million on electricity bills and 620,000 tonnes of CO_2 every year.

Changing behaviour

Individuals can be motivated by fear, greed, fashion, monetary considerations or even guilt. Emotional desires such as the need to protect one's family, or even excitement about saving the world can have an impact. There is one thing, however, that lies at the heart of all motivation and that is information. If you know what's going on and understand the role you have to play, you're far more likely to take action. We all know that we need to cut energy use, cut consumption and cut waste.

Stories abound about people who have managed to cut their carbon footprint, live sustainable lives in some green utopia: they live off the land in renewably powered homes, don't drive or fly and live as close to sustainably as possible. For most of us, who have cars and who buy food at supermarkets, it's not so easy – however hard we try we're made to feel as if we're not quite good enough. But it's surprising how much difference individual action, when measured collectively, can make.

The trouble with changing behaviour is that most of us want things to be easy and to be comfortable. The global recession may well have an effect on energy and product consumption, as there may not be much spare cash available for a TV, iPod, computer, Wii, Nintendo, Blackberry or whatever this week's gadget might be. However, if we're looking to have a genuine impact on the way in which people are going to want to behave over the longer term, we're going to have to address the fundamentals of why and how choices are made in our society.

While as individuals we may feel that we have little effect on large global problems such as climate change, that simply isn't true. We are active consumers in almost everything that we do, and that consumption drives the global markets. The UK's Carbon Trust produced some research identifying that the average UK consumer uses products and services with a combined carbon footprint of 176.4m tonnes carbon (MtC) per annum. This is 11.7 MtC greater than the emissions from all UK production, meaning that the UK is a net importer of carbon intensive products and services from abroad.

UK electricity producers emit almost 24 million tonnes of carbon (MtC) per year, yet this is only 13.6% of the 176.4 MtC embedded in UK products and services. That means that while cutting the emissions from power generation is important, we really need to know more about our individual impact. Managing individual consumption may constitute the area where the greatest mitigation of carbon emissions may be achieved. Only by understanding where carbon is emitted at every point of the supply chain can the average consumer understand how their behaviour affects the amount of carbon consumed. For example, if we look at clothing – emissions from the manufacture of clothing, washing machines and detergent are smaller than from the electricity used in washing, drying and ironing the clothes over their lifetime. So minor changes in the way we wash could even have a major impact.

Taking action

It is behavioural change that will drive down emissions. There are endless sources of information on how to drive down individual fossil-fuel energy use. Obviously onsite renewable power generation is the best, but hardly an option for most consumers. The next best thing is to buy green energy, although that can prove almost as hard. After that, buy a green tariff (although it's worth asking around, as some tariffs are mandated by law and by choosing it, you might well end up paying more with no additional benefit to the energy environment). Turn your appliances off, don't keep them on stand-by – according to figures from the Energy Saving Trust, the UK emits over 3 million tonnes a year of CO_2, just as a result of electrical equipment being left on standby. Domia has even developed a plug-in switch with a remote control that turns devices off at the wall to address this issue.

We know it's sensible to keep the temperature range in the home at a certain level, to change to energy efficient light bulbs, not to waste water. Let's not forget that, according to the US EPA, if you let the tap run for 5 minutes you can use as much energy as a 60 watt bulb running for 14 hours – so don't run the tap while you brush your teeth. Don't use a disposable razor, use a real one; recycle the containers that are accepted in your area; and don't boil a full kettle if you're making one cup of coffee.

Consumers can exercise their power of choice by using recycled goods – recycled paper means less energy and water use, air pollution and waste (but in the US, only accounts for roughly 10% of the market). Walk, cycle or use public transport wherever you can and remember to take shopping bags to the supermarket, don't use plastic. Even something as simple as only doing your washing when you've got a full load in the washing machine can make a contribution.

There are products that can help you keep aware of your power consumption levels, and even turn devices off when they are not being used. One such tracker is the Ambient Orb, a US product, which can be used to illustrate all kinds of data. Connected to a power management platform, it can be set to glow from green to red to illustrate how much power you're using, or when power is cheaper. Southern California Edison used them to help customers change their power consumption patterns. In Europe, the Wattson has also proved popular, which shows how much power each electrical appliance uses when switched on or off, either in watts, pounds or euros. Smart meters that track energy use are also becoming more common. Bye Bye Standby can be used by individuals or offices, providing a smart socket with a group on/off function. A smart plug from AlertMe enables people to manage their appliances remotely over a phone or internet connection.

Waste is a critical issue – around a third of developed world expenditure ends up in the rubbish. It has been estimated that producing a single computer chip, weighing but a few grams, requires 1.7kg of fossil fuel and chemical inputs, as well as 32 litres of water. According to international sustainability expert Annie Leonard, only 1% of purchased goods are still in use 6 months after purchase. Each year in Europe around 7 million tonnes of electronic waste is generated and in spite of the Waste Electrical and Electronic Equipment regulations, much is still finding its way to landfill.

While each individual action may save as little as a few grammes of CO_2, multiplied by hundreds of thousands such actions can make a significant contribution.

Standards

When it comes to using the power of the consumer purse, it's imperative that the right regulations and standards are in place, as these enable consumers to make informed choices about their purchases. Labels that give us information about the relative 'green' nature of our products and services are a key part of the necessary information, but there are so many different standards that it can be difficult to tell what to look for.

Research from Forum for the Future reported that the current and growing mass of environmental claims and ecolabels has confused many consumers and created uncertainty about which claims to trust and how best to make environmentally friendly purchases. Should you buy Fairtrade, organic, green labelled, eco-friendly, free-range, locally sourced, Energy+, Energy Star? Obviously what you should buy will depend on the country, the context and what you're trying to achieve with your own purchasing power.

If you're buying Fairtrade coffee, you're supporting a more equitable and sustainable system of production and trade, intended to benefit people and their communities. If you're buying an EC Energy Labelled fridge, you're probably looking for the most efficient version. If you're buying food with a Soil Association Symbol, you're trying to ensure that you're buying truly organic, pesticide free food. If you're buying products with the Forest Stewardship Council logo, you probably want to know that your packaging, paper or product comes from well-managed timber, as you're concerned about the loss of forests. If you're buying something with the EPA Energy Start Program logo, you're looking for appliances that power down when not in use.

But many of these standards are voluntary, some are difficult to compare to others and others are not precisely defined. In the UK a 'free-range' label doesn't necessarily mean the animals are allowed outdoors or that meat labelled as 'natural' doesn't contain artificial ingredients.

What should we be looking for?

There are a number of key questions that need to be asked when comparing different environmental standards. Firstly, what sort of environmental claim is being made?

Is the manufacturer making a claim about a single environmental attribute such as energy efficiency or recycled-content or is the manufacturer making a broader multi-attribute claim that the product meets an environmental leadership standard? Sometimes even those companies that do have valuable, single-attribute environmental claims fail to address other potentially important human health and environmental issues. Consumers should look for specific statements which can be checked and verified, with as little jargon as possible.

It should be possible to substantiate environmental claims, and obvious what part of the product or process can claim the benefit. In a perfect world, environmental labels would make clear the sustainability implications throughout the lifecycle of the product from raw material extraction and manufacture to use and ultimate disposal.

Next, can you take a look at the details of the environmental standard in question, or at the testing protocol? If a manufacturer can not or refuses to provide a copy of this, one might suspect that the claim for the product is mainly a marketing ploy. When a copy of the standard is provided, make sure it refers to the appropriate national or international environmental and performance standards.

Finally, try to find out how the standard was developed. You really want to know that it was developed in a public, transparent process similar to the way ANSI, ASTM, ISO 14024 or other public standards are developed. The most trusted ones are usually those developed in a consensus-based process by broad stakeholder groups. Standards developed consistently with ISO 14024 protocols will make a list of stakeholder groups available upon request. Consumers should be less trustful of standards developed by an individual manufacturer or trade association because of potentially unmitigated conflicts of interest.

Consumers and greenwash

Greenwash is a major problem for the consumer. It can provide confused or reluctant customers with an excuse to do nothing. The problem here is that when dealing with environmental issues, companies don't speak to consumers in a language they understand, undermining the potential for positive green behaviour.

UK-headquartered sustainable communications group Futerra believes that most greenwash is due to ignorance or sloppiness rather than deliberate intent. In 2008 it released *The Greenwash Guide* to help companies work out what to avoid. It also helps outline what consumers should look out for:

- Fluffy language: words or terms with no clear meaning, e.g. 'ecofriendly'.
- Green products vs dirty company – such as efficient light bulbs made in a factory which pollutes rivers.
- Suggestive pictures: green images that indicate an (unjustified) green impact, e.g. flowers blooming from exhaust pipes.
- Irrelevant claims: emphasising one tiny green attribute when 'business as usual' isn't green.
- Best in a bad class?: this is where the company is slightly greener than the rest, even if the rest are pretty terrible.
- When it's just not credible: 'ecofriendly' cigarettes anyone?

'Greening' a dangerous product doesn't make it safe.
- Gobbledygook: jargon and information that only a scientist could check or understand.
- Imaginary friends: a 'label' that looks like third-party endorsement ... except it is made up by the company itself.
- No proof: it could be right, but where's the evidence?
- Outright lying: totally fabricated claims or data.*

*The Greenwash Guide, Futerra Sustainability Communications, May 2008

What's critical here is that consumers begin to demand full and accurate information. We need companies to be specific about their actions. We don't want them to use jargon, but just to tell us what they're doing and why. And we should encourage everyone we know to stop using terms that don't really mean anything without the necessary information. Every time we see terms like 'green', 'carbon neutral', 'environmentally friendly', even 'energy efficient' (unless we've got standards and benchmarks for comparison) we should demand a detailed explanation. In that way we can train companies, and marketers, as to what we're prepared to accept.

Cultural change

There has been much discussion about whether it is no longer 'fashionable' to be green and about whether or not the coming recession is going to put taking green action at the bottom of the individual agenda. Focus on the green agenda, however, is about far more than fashion. There seems to be a growing awareness of the fact that the way in which our societies function may not be the best, environmentally, economically or even in terms of individual quality of life.

There is a growing sense that we must begin to take responsibility for the long-term consequences of our actions and that this is to be lauded and respected. When Greenpeace protestors took action at the site of the new Kingsnorth coal-fired power station, they were arrested and many people assumed that they would be punished for criminal damage. At the 2008 trial, however, they used the defence of a 'lawful excuse', arguing that taking action to prevent climate change was a reasonable step. And they won. This was the first time that a UK court had allowed such a defence but it may not be the last.

It's time that we as individuals take action on our own behalf to create the world in which we can, and want to live. The global economic crisis resulted in many people attacking the banking system, as well as bankers themselves, blaming them for our economic woes. While it's probable that we all put too much faith in 'experts', it's also true that as long as things were going well, few people bothered to question the system. It's about time that changed.

The transition movement: taking action collectively

The zeitgeist certainly seems to be changing and for many individuals the key question is now one of individual responsibility and how we can take action in a collective way. Cultural changes begin at different levels. An example of how different strands can entwine is the Transition Town movement. This was founded in 2006, in response to concerns about the twin problems of climate change and peak oil.

Transition is the brainchild of Rob Hopkins, who started Transition Totnes (the first experiment in developing a transition community) in 2006. Once oil supplies peak, at some point there won't be sufficient oil to meet demand and then we'll have to get along without it. He compares the concept of peak oil to climate change in his 2006 book, The Transition Handbook, by saying that whereas the dangers of climate change say that we should change, the consequences of peak oil say that we'll be forced to change.

The idea behind transition is that we need to prepare for a potential energy crisis. Central to the success of transition is the notion that, as individuals and communities, we can cut carbon and build resilience in our local communities in order to withstand change. This will mean an increase in local energy, local food and local finance, as well as reskilling. A village, town or city needs to be able to depend on its own resources: the more food, power and other necessities you can produce in your area, the less you rely on imports.

Critical to the concept of transition is an energy descent plan – a Plan B for what to do when cheap energy isn't easily available. Transition teams can work with local business on oil vulnerability analysis – basically asking if we can keep machines running. Hopkins points out that when some ecosystems are hit they collapse, while others survive. It's likely to prove similar in towns and communities, where there is little to no local ownership, no local food sources. Once you start to withdraw liquid fuel, there's no alternative to localization.

Food security and supply is an interesting problem. If global supply chains dwindled, where would you get your food? During the Second World War, between 1939 and 1944, food imports to Britain halved and the country nearly doubled domestic food production. Back then London's market gardens alone produced 45 tonnes a hectare but today there is comparatively little food grown in a city that feeds at least 7 million people every day. What would happen if there were problems with the food supply?

Following the implementation of a US trade embargo (which began in 1962) Cuba began to suffer food supply shortages. When this was followed by the collapse of the Soviet bloc, the country lost access to cheap fossil fuel, food imports and agricultural inputs. During the early to mid-1990s, known as the 'special period', food rationing was introduced in Cuba to deal with the problem. Cuban leader Fidel Castro declared that no land should be left uncultivated and urban food gardens became a critical part of feeding the people. This local food self-sufficiency also reduced the need for transport, refrigeration and storage. Today Havana's urban gardens account for around 250 tonnes of produce per hectare and are a prime example of taking the community approach.

But it's also more than that – if we can see the coming changes as an opportunity then we're open to Transition. Hopkins defines the essence of Transition as the idea that 'the future with less oil could be preferable to the present'.

Transition recommends getting involved locally as developers, ensuring that new developments are kept radically low carbon. It also underscores the need to think about embedded carbon in design, planning and construction. Working together as a community also builds up procurement power – as a group negotiating deals on solar panels would be much more effective. In Ouse Valley in the UK, they went one stage further and set up their own energy services company, Ovesco. It's vital to find ways to invest money into the community.

Local finance is a key element of Transition, with many communities exploring the impact of their own currency. According to analysis from the New Economic Foundation, if you shop in mainstream shops, around 80% of the money leaves the area. If you shop at local businesses, the proportion is reversed with 80% of that money staying in the area – a currency that only operates locally is a powerful nudge towards keeping the majority of the community's funds within that community and supporting local development.

There are several other principles underlying Transition: these include providing a positive vision rather than making people afraid of what might be on the way; helping people to access good accurate information and trusting them to make good decisions. Inclusion and openness are key, the scale of what is needed is enormous and must include environmentalists, business, faith groups, schools, as well as individuals enabling change, sharing methodology and best practice.

Transition itself is intended to be a catalyst for communities which want to take control of their own environment. The movement makes recommendations and provides support through the Transition Network, but it's up to individuals to move as fast, or slowly, as they would like. Transition does recommend things: planting nut trees which have more protein and carbohydrate than cereals; creating local food directories to link local growers, consumers and retailers; printing local money as part of a new story about money and ownership; sharing gardens; retraining – learning to grow, fix and build things.

The movement is gaining in popularity, with the 100th formal transition project set up in Fujima, Japan, in 2008. There are transition projects in the US, New Zealand and Australia, as well as the UK, and over 900 communities mulling the prospect of adopting its principles. In effect the transition movement is a call to our adaptability and resilience, our ability to imagine a better way of living. It's important to recognize that what we're dealing with is no longer a single issue, but it's about power, water, food, creativity and more – it's about giving power back to individuals to choose how to live their lives.

Consumers clearly have a critical role to play in demanding information, supporting positive actions either from corporations or politicians but, most importantly, in taking collective action which supports the transition towards a low carbon economy.

Conclusion

Conquering carbon is a critical step in the evolution of a global society. If we can find a way to create an equitable, effective and economic process for controlling our emissions, it could be the beginning of a re-evaluation of our relationship with the world around us. Climate change has the potential to change all of our lives in negative ways. If we want to balance the odds in our favour and minimize the chance of rising sea levels, increasing droughts, and difficulties in accessing food, water and health care, we need to act and we need to act now.

It's obvious that climate change will diminish our ability to provide sufficient resources to support the growing global population. While some believe that the only solution to this problem is to control population, controlling consumption might prove just as effective. Current global population growth is around 1.2% while the economic growth rate is around 3.8% – that means that consumption is responsible for two thirds of the economic growth rate. If we are able to address this issue, we could begin to address related environmental issues. That doesn't mean that we need to return to subsistence living but it does mean that we need to take a fresh look at our expectations of how we should live.

We need to rethink how we value the world and, once we've understood the impact of our consumption, price that impact into day-to-day life. Economic and climate modelling are a necessary part of future climate change negotiations, in order to support the most effective and economically viable technologies and policies to reduce GHG emissions. If we're going to get action there needs to be a sense of urgency so that change can be included in the planning process, for governments, for cities, for corporations and for consumers.

The mitigation of climate change poses its own set of challenges and opportunities, including changes in consumer demand for products and services, compliance with regulation, new risks to brand value and changes in shareholder demands. A combination of measures and policy tools needs to be implemented to provide the right environment for change. These have to be targeted at all life-cycle stages and to all major stakeholders. How we respond to the threat of climate change is not a technological challenge, it's a political, financial and cultural one. We need action on the international stage and we need to review the tenets of our society – what on earth is the point of consumption for its own sake?

We need to implement the necessary measures to cut carbon from our economy, whether through taxes, standards and regulation or through emissions trading. The success of the US sulphur market, cuts in global ozone emissions and the impact of the collective action that led to the fall of the Berlin Wall in 1989 are all examples of what we can achieve once we decide to act together. Change achieved through direct action or change achieved through diplomatic negotiations sometimes need to go hand in hand.

The role of the carbon markets

The major benefit of an effective carbon market is that it provides a model that allows the markets to find the cheapest and most efficient option to achieve targeted emissions cuts. The EU ETS has suffered some teething problems mostly due to decisions made on the basis of lobbying from industry and individual governments. Lessons have been learned which, if implemented, could ensure that the carbon markets could provide a environment rich with both risk and opportunity, and accelerate the private sector's acceptance of the needed transition to a low carbon economy.

Another vital role that the carbon markets can play is in the ability of the markets to raise funds for action. The proposed US cap-and-trade market is intended to pay for investments into green energy, infrastructure upgrades and more. The US RGGI market raised over $171 million in its first compliance auction in March 2009 which will be used to pay for energy efficiency, renewable energy and other consumer benefit programs in the 10 RGGI states (Connecticut, Delaware, Maine, Maryland, Massachusetts, New Hampshire, New Jersey, New York, Rhode Island and Vermont).

It's possible that the notion of a well-regulated market with public good as an end goal could help restore public trust in the financial markets, while stimulating the economy at the same time. If we focus our efforts on driving forward markets regulated for a public good, with the market confidence that this would support, we could soon be well on the way to economic recovery.

We want to see a global climate change and carbon agreement so that governments, banks, entrepreneurs and industry all have a clear picture of where the market is going. This is one of the drivers behind the attempt to reach a post-Kyoto agreement at Copenhagen in 2009. We may be able to achieve change without it but it's likely to prove harder. It's possible, however, that if the economic imperative shows that industry will become more competitive through the implementation of energy efficiency measures, and more aware of the opportunities provided by new technology sectors, then the private sector could well continue down a low carbon path.

More importantly, we may be able to achieve an international carbon price outside a post-Kyoto treaty. It's clear that we are looking forward to a world which has a plurality of carbon markets. The EU, the US, Australia, New Zealand, Japan and China already have, or are planning to develop, markets to trade carbon. It's possible that, with the experience we have in connecting different markets through finance (as for example in the development of the global currency markets) we might be taking the first steps towards a global carbon price, and a global incentive for all economies to change direction.

Conclusion

Of course, we are more likely to succeed in a transition to a low carbon economy within the framework of an international treaty. We look to the international community to set limits, timeframes and mechanisms for the removal of carbon from the global economy. There are reforms that could be made in a post-Kyoto climate change agreement, be they improvements on the existing system or a reworking of the underlying framework of Kyoto.

These could include the full introduction of forestry into the CDM. Paying developing countries for retaining their standing forests might prove one way to solve the dilemma of how the developed world could fund mitigation and adaptation in the developing world. Tightening the global emissions regime could compensate for the addition of new credits to a global trading scheme.

At the same time, introducing emissions targets for industry sectors could ensure that emissions reductions become the responsibility of industry, easing national concerns about competitiveness. This would work most effectively for those industries which, by definition, cross international boundaries, such as aviation and shipping. The EU is a strong proponent of this option, given the range of national wealth within the EU itself, focusing on the idea of targets for all but differentiated levels of responsibility.

Whatever route the international negotiations take, there are a few areas upon which most people are agreed. Any post-Kyoto deal should support:

- Investment in infrastructure
- A global carbon cap, diminishing annually to 2050
- Better information and communication about GHGs and their emission and impact
- A global agreement which enables different countries to take the approach most suited to their needs
- Mechanisms for the reduction of GHG emissions
- Mechanisms for the funding of mitigation of and adaptation to the impact of climate change

Significant barriers in the negotiations for a post-Kyoto treaty remain. Firstly we have to agree on a global emissions cap to which all parties can contribute. The IPCC recommendation ranges between 60–80% cuts from 1990 levels. Given that the science suggests that this is merely cutting the odds of undergoing dramatic climate change, it might be best to set a goal for that higher target.

While the US has demanded limits to growth in developing world emissions as a pre-requisite for an international agreement, China and India have argued strongly for their right to independent economic development. In order to resolve the question of whether emissions limits should be set as absolutes, or on a per capita basis, perhaps we could set a formula which uses the two? Initial caps set based on current emissions, translating to a per capita basis over the next 10 to 20 years.

Payment for mitigation and adaptation is critical, and current funding propositions fall well below the finance required to have any real impact. If we set a global financial cap, under which every party is allocated emissions, those emissions allowances should be auctioned. Revenues from those auctions could be split between a percentage spent on climate change projects in each country, and a percentage paid into a global mitigation and adaptation fund.

Beyond this, we also need policy support for increased energy efficiency standards, fuel consumption requirements; clean energy and other clean technologies. These can happen either inside or outside an international climate change agreement, however.

Individual countries need to set up strong frameworks for the deployment of renewable power generation; electric vehicles to work in conjunction with renewables to provide energy storage; increased decentralization of power generation; the development of a Smart Grid to replace the current transmission and distribution networks; a Super Grid to work in conjunction with localized generation. We need support for innovation and new technological breakthroughs.

We'll need new financial frameworks as well, such as the set up of loan guarantee schemes. This could enable companies and consumers to pay for the implementation of new technologies, energy solutions and efficiencies, and pay the loan back out of future savings.

The policy shift following US President Barack Obama's election has brought new hope to international climate change negotiations. If a federal cap-and-trade scheme is being deployed specifically to address the dangers of climate change, the likelihood of an agreement on a post-Kyoto Protocol that includes the US seems, for the first time, a very real possibility.

The three key economic regions that must take part in a global agreement if it has any chance of having an effect are the EU, US and China. While the EU is already committed to long term caps, there is sufficient uncertainty about the position of the US and China that it's questionable whether a final agreement can be negotiated within the time frame set by the Bali Roadmap. Obama's move to get approval of a US federal carbon scheme through the Budget was clever, but it's possible that it will take a significant amount of time to set up the legal framework for such a scheme. Wrangling over the right approach to take, especially in light of the general economic malaise and the US involvement in overseas conflicts, could also slow the process. But there is no question that the US position has fundamentally changed with the new administration. There is equally no doubt that China is moving forward with its own plans to control emissions, with a number of exchanges launching with plans for trading in credits for SO_2, CO_2 and water. There is clearly hope for compromise.

It's worth noting that, according to analysis by IDEA carbon, the EU's target of a 20% reduction from 1990 levels by 2020 is equivalent to a 14% reduction from 2005 levels. This means that the new US target is, in fact, not that different from that of the EU (although a reduction to 1990 levels by 2020 for the US would be equivalent to a 16% cut from 2005 levels). Effectively, the US has set a goal that is consistent with EU targets. Critically, the introduction of such a scheme in the US, with targets that suggest a commonality with the EU approach, gives the first concrete hope of a global carbon market.

One that could potentially develop and survive even beyond the framework of a post-Kyoto deal. The critical need to address the imbalance of supply and demand in diminishing global resources means that there is increasing pressure on members of the international community to take action unilaterally.

A chance at a more equitable world

Whatever the result of international climate negotiations, individuals are waking up to the idea that they have a responsibility to the global environment. There is a recognition that collective action – be it through the use of your wallet, your vote or just in taking a stand by boycotting a particular good or service – can make a difference. If change is to be effected, it must take place at every level of the economy and every level of society.

Everyone has a carbon footprint and there are options to enable individuals to offset and lower their emissions. Business support for the development of carbon-neutral products and services could provide a way to remove carbon from the supply chain where possible. Providing an opportunity for development rather than legislative regulation could have a dramatic effect – if consumers used their buying power to affect the development of carbon-neutral products and services, rapid market change could be effected.

The transition to a low carbon economy will result in the biggest technological and economic transformation since the Industrial Revolution. It's a gamble, possibly the biggest humankind has ever faced but we should be able to achieve our goals. While the carbon markets are not the only tool we can use, they offer an opportunity to transform our economic framework and bring a connection between rights and responsibilities back into the marketplace.

We have the means to change the way things work if we understand the stakes, implement the mechanisms and make the necessary decision to act. If we take this opportunity to transform the way in which we value our world, we might save the economy, the planet and ourselves.

Further Reading

Introduction

IPCC 2007, *Climate Change 2007: The Physical Science Basis. Contribution of Working Group I to the Fourth Assessment Report of the Intergovernmental Panel on Climate Change*, Solomon, S., D. Qin, M. Manning, Z. Chen, M. Marquis, K.B. Averyt, M.Tignor and H.L. Miller (eds.). Cambridge University Press, Cambridge, United Kingdom and New York, NY, USA.

IPCC 2007, *Climate Change 2007 – Impacts, Adaptation and Vulnerability. Contribution of Working Group II to the Fourth Assessment Report of the Intergovernmental Panel on Climate Change*, Parry, M. Canziani, O. Palutikof, J. van der Linden, P. Hanson, C. (eds)]. Cambridge University Press, Cambridge, United Kingdom and New York, NY, USA.

IPCC 2007, *Climate Change 2007 – Mitigation of Climate Change. Contribution of Working Group III to the Fourth Assessment Report of the IPCC*, Metz, B. Davidson, O. Bosch, P. Dave, R. Meyer, L. (eds), Cambridge University Press, Cambridge, United Kingdom and New York, NY, USA.

IPCC, 2007, *Climate Change 2007: Synthesis Report. Summary for Policymakers*, Lenny Bernstein, Peter Bosch, Osvaldo Canziani, Zhenlin Chen, Renate Christ, Ogunlade Davidson, William Hare, Saleemul Huq, David Karoly, Vladimir Kattsov, Zbigniew Kundzewicz, Jian Liu, Ulrike Lohmann, Martin Manning, Taroh Matsuno, Bettina Menne, Bert Metz, Monirul Mirza, Neville Nicholls, Leonard Nurse, Rajendra Pachauri, Jean Palutikof, Martin Parry, Dahe Qin, Nijavalli Ravindranath, Andy Reisinger, Jiawen Ren, Keywan Riahi, Cynthia Rosenzweig, Matilde Rusticucci, Stephen Schneider, Youba Sokona, Susan Solomon, Peter Stott, Ronald Stouffer, Taishi Sugiyama, Rob Swart, Dennis Tirpak, Coleen Vogel, Gary Yohe, 2007. Cambridge University Press, Cambridge, United Kingdom and New York, NY, USA.

United Nations, World Population Prospects: The 2008 Revision Population Database.

International Energy Agency, Energy Technology Perspectives 2008 – Scenarios and Strategies to 2050, 2008

Chapter 1: Climate Change and Greenhouse Gases

UNEP Year Book 2009, *New Science and Developments in our Changing Environment*, United Nations Environment Programme.

Climate Change, *Global Risks, Challenges & Decisions*, 2009, University of Copenhagen.

Dr. James Hansen et al, *Target Atmospheric CO2: Where Should Humanity Aim?*, The Open Atmospheric Science Journal, November 2008

Emissions of Greenhouse Gases Report, December 3, 2008, US Energy Information Administration

Stern, N., et al, HM Treasury, *Stern Review: The Economics of Climate Change*, 2006.

Tol, R. S. J., Economic and Social Research Institute, Hamburg, *The Stern Review of the Economics of Climate Change: A comment*, November 2006.

Fankhauser, S., and Tol, R. S. J., Resource and Energy Economics, *On Climate Change and Economic Growth*, 2005.

Hope, C. W., Integrated Assessment Journal, *The Marginal Impact of CO2 from PAGE2002: An Integrated Assessment Model Incorporating the IPCC's Five Reasons for Concern*, 2006.

Novatlantis, *Smarter Living. Generating a new*

understanding for natural resources as the key to sustainable development – the 2000-watt society, March 2005.

Australian Treasury and Ministry for Climate Change and Water, Australia's Low Pollution Future: The Economics of Climate Change Mitigation, 30 October 2008.

Chapter 2: The Resource Wars

UN Environment Programme (UNEP)'s Year Book 2009, New Science and Developments in our Changing Environment, United Nations Environment Programme.

UNEP's Post Conflict Environmental Assessment of Sudan, June 2007,

Fankhauser et al, Valuing Climate Change: The Economics of the Greenhouse Effect, 1995

Climate Change – the Costs of Inaction, Frank Ackerman and Elizabeth Stanton, Global Development and Environment Institute, Tufts University

WWF. The Living Planet Report 2008, 2008.

Environment a Growing Driver in Displacement of People, Michael Renner, September 2008, The Worldwatch Institute, Vital Signs 2007-2008: The Trends that are Shaping Our Future, 2008.

United Nations University, Overcoming one of the greatest environmental challenges of our times: rethinking policies to cope with desertification, 2007.

The Worldwatch Institute, State of the World 2008: Toward a Sustainable Global Economy, 2008.

United Nations Food and Agricultural Organisation, FAO Food Outlook, May 2008,

UN FAO, Crop Prospects and Food Situation, No 1, February 2008, Global Information and Early Warning System.

IPCC June 2008, Technical Paper VI, Climate Change and Water, Bates, B.C., Z.W. Kundzewicz, S. Wu and J.P. Palutikof, (eds).

McGranahan, G., D. Balk, and B. Anderson. Environment & Urbanization, The rising tide: assessing the risks of climate change and human settlements in low elevation coastal zones., 2007.

Organisation for Economic co-operation and Development (OECD). Agrawala, S. Ota, T. Ahmed, A.U. Smith, J. van Aalst, M., Development and Climate Change in Bangladesh: Focus on Coastal flooding and the Sundarbans, 2003.

Prepared by Science Applications International Corporation (SAIC) for the US Department of Energy. Hirsch, R. L. et al, Peaking of World Oil Production: Impacts, Mitigation and Risk Management, February 2005.

Strahan, D., John Murray, The Last Oil Shock: A Survival Guide to the Imminent Extinction of Petroleum Man, 2007.

International Food Policy Research Institute, von Braun, J., The World Food Situation: New Driving Forces and Required Actions, December 2007.

Proceedings of the National Academy of Science (PNAS), Tubiello, F. N., Soussana, J-F., Howden, S.M., Crop and pasture response to climate change, December 2007.

PNAS, Tubiello, F. N., Soussana, J-F., Howden, S.M. Chetri, N. Dunlop, M, Meinke, H., Adapting agriculture to climate change, December 2007.

PNAS, Schmidhuber, J., Tubiello, F. N., Global food security under climate change, December 2007.

Cline, W.R. Center for Global Development. Global Warming and Agriculture: Impact Estimates by Country, July 2007.

United Nations, World Water Development Report 3. Water in a Changing World, March 2009.

Hoekstra, A. Y., and Chapagain, A. K., Globalization of water: Sharing the planet's freshwater resources, Blackwell Publishing, 2008.

Further Reading

Global Water Partnership, *Planning for a water secure future: Lessons from water management planning in Africa*, 2006

WWF, Pittock, J., *Water for life: Lessons for climate change adaptation from better management of rivers for people and nature.*

WWF., *Adapting water to a changing climate: An overview*, August 20008.

California Energy Commission, Krebs, M., *Water-Related Energy Use in California*, February 2007.

European Union, Sukdev P., *The Economics of Ecosystems and Biodiversity (TEEB), Interim Report*, European Communities May 2008.

Chapter 3: Driving Change

Vattenfall, *Global Mapping of Greenhouse Gas Abatement Opportunities*, January 2007.

Policy Exchange, Spracklen, D., Yaron, G., Singh, T., Righelato, R., Sweetman, T., (ed. Caldecott, B.,), *The Root of the Matter*, August 2008.

New Energy Finance and UNEP, *Global Trends in Sustainable Investment 2008.*

World Economic Forum, *Green Investing. Towards a Clean Energy Infrastructure*, January 2009.

UNEP Green Economy Report, *Global Green New Deal* (part of the Green Economy Initiative), October 2008

Green New Deal Group, New Economics Foundation, *Green New Deal*, July 2008.

Forum for the Future, *Getting to Zero: Defining Corporate Carbon Neutrality*, June 2008.

Carbon Trust, *Cutting Carbon in Europe. The 2020 plans and the future of the EU ETS*, June 2008.

Weber, C.L., *Energy Policy*, July 2008.

Matthews, H. S., Hendrickson, C., Weber, C., *Environmental Science & Technology, The Importance of Carbon Footprint Estimation Boundaries*, 2008.

Carbon Trust, *Carbon footprinting. An Introduction for Organisations*, August 2007.

Carbon Trust, *Product carbon footprinting: the new business opportunity*, October 2008.

Bhatia, P., Ranganathan, J., World Business Council for Sustainable Development and the World Resources Institute, *The Greenhouse Gas Protocol: A Corporate Accounting and Reporting Standard*, (Revised Edition), March 2004.

Carbon Trust, *Code of Good Practice for Product Greenhouse Gas Emissions and Reduction Claims*, October 2008.

BSI Standards, *PAS 2050:2008 – Specification for the assessment of the life cycle greenhouse gas emissions of goods and services*, 2008

Chapter 4: Decarbonizing the World

International Energy Agency (IEA), *Energy Technology Perspectives 2008 – Scenarios and Strategies to 2050*, 2008

'Sustainable Development International in partnership with the United Nations Environment Programme', *Climate Action*, 2007–2008, 2008

Goodall, C., *Green Profile, Ten Technologies to Save the Planet*, 2008.

Socolow, R., Pacala, S., *Science, Stablization Wedges: Solving the Climate for the Next 50 Years with Current Technologies*, August 2004.

IEA, *World Energy Outlook 2008*, November 2008.

Google, *Clean Energy 2030: Google's Proposal for reducing US dependence on fossil fuels*, October 2008.

Greenpeace and the European Renewable Energy Council (EREC), *Energy [r]evolution: A sustainable global energy outlook*, 2008

McKinsey Global Institute, *The case for investing in energy productivity*, February 2008.

McKinsey Global Institute, *The carbon productivity challenge: Curbing climate change and sustaining economic growth*, June 2008.

Lovins, A. B., Sheikh, I., Markevich, A., *Nuclear Power: Climate Fix or Folly?*, RMI Solutions article *Forget Nuclear*, updated and expanded by ABL 31 Dec 2008. Originally published, April 2008.

Greenpeace, *The True Cost of Coal*, November 2008.

Naucler, T., Campbell, W., Ruijs, J., McKinsey & Company, *Carbon capture & storage: Assessing the economics*, 2008.

Archer, C.L., Jacobson, M. Z., Global Climate and Energy Project (GCEP), Stanford University. Published in the Journal of Geophysical Research – Atmosphere 2005, *Evaluation of global wind power*

Germany Advisory Council on Global Change. (WBGU), *World in Transition: Towards Sustainable Energy Systems*, 2003.

Institute for Energy and Environmental Research (IEER), *Cash Crop on the Wind Farm*, April 2004.

North American Electric Reliability Corporation (NERC), *2008 Long-Term Reliability Assessment*, 2008.

Global Energy Network Institute (GENI), *Global Integrated Energy Models: a summary*, Ongoing.

German Federal Ministry for the Environment, Nature Conservation and Nuclear Safety (BMU), Desertec, MED-SP: *Concentrating Solar Power for the Mediterranean Region*, 2005.

German Federal Ministry for the Environment, Nature Conservation and Nuclear Safety (BMU), Desertec, TRANS-CSP *Trans-Mediterranean interconnection for Concentrating Solar Power*, 2006

German Federal Ministry for the Environment, Nature Conservation and Nuclear Safety (BMU), Desertec, *Concentrating Solar Power for Seawater Desalination*, 2007.

Institute for Energy, Distributed Power Generation in Europe: technical issues for further integration (Angelo L'Abbate, Gianluca Fulli, Fred Starr, Stathis D. Peteves), 2008.

McKinsey Global Institute, *Averting the next energy crisis: The demand challenge*, March 2009.

National Energy Technology Laboratory (NETL), *A Vision for the Modern Grid. The Modern Grid Initiative*, August 2008.

Boccaletti, G., Loffler, M., Oppenheim, J. M., McKinsey Global Institute, *How IT Can Cut Carbon Emissions*, October 2008.

McKinsey & Company, *Pathways to a Low Carbon Economy*, January 2009. (update of McKinsey report, Greenhouse Gas Emissions Abatement Cost Curve)

Romm, J., RAND Corporation, *The Internet and the New Energy Economy*.

Mazza, P., Pacific Northwest National Laboratory (PNNL), *The Smart Energy Network: Electrical Power for the 21st Century*, 2006

McKinsey Global Institute, *Curbing global energy demand growth: The energy productivity opportunity*, May 2007.

Gartner, *Emerging Trends*, Gartner Symposium/IT Expo 2007.

The Climate Group, SMART2020: *Enabling the low carbon economy in the information age*, June 2008.

UNEP, *Buildings and Climate Change – Status, Challenges and Opportunities*, 2007.

European Telecommunications Network Operators Association (ETNO) and WWF, *Saving the Climate @ the Speed of Light: First roadmap for reduced CO2 emissions in the EU and beyond*, 2006.

Scott, F., Green Alliance, *Teaching homes to be green: smart homes and the environment*, November 2007.

Lorch, R., (ed), Building Research & Information, *Climate Change: National Building Stocks*, 2007.

Further Reading

Shove et al., Lorch, R., Building Research & Information, Comfort in a Lower Carbon Society, July 2008.

Berntsen, T., Center for International climate and Environmental Research (CICERO), The effective of transport emissions on the climate, October 2004.

Torvanger, Asbjørn, Bjørg Bogstrand, Ragnhild Bieltvedt Skeie and Jan S. Fuglestvedt, CICERO, Climate regulation of ships, July 2007.

Ernst & Young, Inclusion of Aviation in the EU ETS: Cases for Carbon Leakage, November 2008.

International Association of Independent Tanker Owners (Intertanko).

Doornbosch, R., Steenblik, R., Organisation for Economic Co-operation and Development (OECD), Biofuels: Is the Cure Worse than the Disease, September 2007.

OECD, Policy Brief, Biofuels for Transport: Policies and Possibilities, November 2007.

OECD, Developments in Bioenergy Production across the World – Electricity, Heat and second generation Biofuels, November 2008.

OECD, Energy for Sustainable Development, 2007.

Safadi, R., Trade and Agriculture Directorate, OECD, Rising Food Prices, Causes and Consequences, 2008.

OECD, IMG, World Bank, Workshop on Food and Fuel Prices, Summary of Workshop, September 2008.

New Energy Finance, Food price increases: is it fair to blame biofuels?, May 2008.

OECD, Climate change mitigation: What do we do?, 2008.

IEA, CO2 Emissions from Fuel Combustion, 2008 Edition, 2008

Royal Society, Sustainable Biofuels: Prospects and challenges, January 2008.

Renewable Fuels Agency, Gallagher Reivew of the Indirect Effects of Biofuels, March 2008.

Arup and Cenex, for BERR, Investigation into the Scope for the Transport Sector to Switch to Electric Vehicles and Plug-in Hybrid Vehicles, October 2008.

Penner, J. E., Lister, D. H., Griggs, D. J., Dokken, D. J., McFarland, M., IPCC, Special Report on Aviation and the Global Atmosphere, 1999.

International Maritime Organisation (IMO), Update of the 2000 IMO Study on GHG Emissions from Ships, 2008.

Davieet, F., Greenhalgh, S., Weninger, E., World Resources Institute (WRI), The Land Use, Land-Use Change, and Forestry Guidance for Greenhouse Gas Project Accounting, October 2006.

Millenium Ecosystem Assessment, Millenium Assessment (MA) Board, Living Beyond our Means: Natural Assetss and Human Well-Being, MA Board Statement and summary of the Reports, March 2005.

Island Press, Millenium Ecosystesm Assessment Toolkit, 2007

World Resources Institute, Beyond Carbon Financing, November 2008.

Brown, D., Bird, N., Overseas Development Institute (ODI), The REDD road to Cophenhagen: Readiness for what?, December 2008.

Streck, C., O'Sullivan, R., Janson-Smith, T., Tarasofsky, R., (eds). Brookings Institute, Climate Change and Forests: Emerging Policy and Market Opportunities, May 2008

Spraclen, D., Yaron, G., Singh, T., Righelato, R., Sweetman, T., Caldecott, B., (ed) Policy Exchange, Root of the Matter: Carbon Sequestions in Forests and Peatlands, 2008.

Nobre, A., Instituto Nacional de Pesquisas da Amazonia (INPA), International Institute for Sustainable Development presentation, March 2006.

Ebeling, J., Oxford University MSC dissertation, Tropical Deforestation and

Climate Change: Towards an International Mitigation Strategy, August 2006.

Ebeling, J., Yasue, M., Generating carbon finance through avoided deforestation and its potential to create climatic, conservation and human development benefits, Philosophical Transactions of the Royal Society, 2008.

Bozmoski, A., Hepburn, C., Smith School of Enterprise and the Environment, The Interminable Politics of Forest Carbon: an EU Outlook, 2008.

Fry, I., Review of European Community & International Environmental Law, Reducing Emissions from Deforestation and Forest Degradation: Opportunities and Pitfalls in Developing a New Legal Regime, 2008

Tomaselli, I., United Nations Forum on Forests, Brief study on funding and finance for forestry and forest-based sector, 2006.

UNFCCC (2001) Decision 11/CP.7, Land use, land-use change and forestry (10 November).

UNFCCC (2007) Bali Action Plan, Decision 1/CP.13.

Chapter 5: The International Policy Framework

UNFCCC, Understanding Climate change: A beginners guide to the UN Framework Convention and its Kyoto Protocol, 2002.

New Energy Finance, How to save the planet: be Nice, Retaliatory, Forgiving & Clear, September 2007

New Energy Finance, Towards a workable post-Kyoto international framework for emissions reduction, December 2005.

New Energy Finance Davos Fact Pack, Global Trends in Clean Energy Investment, January 2009.

ILEX report for WWF, The environmental effectiveness of the EU ETS: analysis of caps, November 2005.

Ellerman, A. D., Joskow, P. L., Prepared for the Pew Center on Global Climate Change, The

European Union's Emissions Trading System in Perspective, May 2008.

Burton, I., Diringer, E., Smith, J., Prepared for the Pew Center on Global Climate Change, Adaptation to Climate Change: International Policy Options, November 2006.

New Carbon Finance, Report for WWF, The impact of auctioning on European wholesale electricity prices post-2012, September 2008.

Point Carbon report for WWF, EU ETS Phase II – The potential and scale of windfall profits in the power sector, March 2008.

European Commission, Proposal for a Directive of the European Parliament and of the Council on the promotion of the use of energy from renewable sources, January 2008.

World Bank, States and Trends of the Carbon Market 2008, May 2008.

Chapple, A., Forum for the Future, Making the voluntary carbon market work for the poor, July 2008.

Helme, N., Schmidt, J., Creating Incentives to Reduce Greenhouse Gas Emissions post 2012: Options from the Future Actions Dialogue, November 2006.

Centre for Clean Air Policy, Sectoral Approaches: A pathway to nationally appropriate mitigation actions, December 2008.

Meyer, A., Schumacher Briefing, Contraction & Convergence. The Global Solution to Climate Change, December 2000.

Starkey, R., Anderson, K., Tyndall Centre for Climate Change Research, Domestic Tradable Quotas: A policy instrument for reducing greenhouse gas emissions from energy use, December 2005.

Zanni, A., Bristow, A., Wardman, M., Personal carbon trading and carbon tax; exploring behavioural response in personal transport and domestic energy use, 2008.

Chapter 6: What We Can Do

Honghong Yi, Energy Policy Volume 35, issue

Further Reading

2, Atmospheric environmental protection in China: Current status, developmental trend and research emphasis, February 2007.

Climate Group, China's Clean Revolution, 2008.

Chinese government's second white paper on climate change, Xie Zhenhua, vice-director of the National Reform and Development Commission (NDRC)

Greater London Authority, Action Today to Protect tomorrow: The Mayors Climate Change Action Plan, February 2007.

World Bank, Latin American and Caribbean Region Sustainable Development Working Paper 19, Responding to Climate Change: Proposed Action Plan for the World Bank in Latin America, June 2004.

San Francisco Department of the Environment and Public Utilities Commission, Climate Action Plan for San Francisco: Local actions to reduce greenhouse gas emissions, September 2004.

Seattle Climate Action Plan, Seattle, a Climate of Change: Meeting the Kyoto Challenge, September 2006.

Worldwatch Institute, State of the World 2007: Our Urban Future, China and Rizhao, November 2006.

The Climate Group, Consumers, Brands and Climate change 2008 (UK), 2008.

Terra Choice, 2009, The Seven Sins of Greenwashing: Environmental Claims in Consumer Markets, Upgraded from the 2007 report, The Six Sins of Greenwashing

Grant, J., Wiley, The Green Marketing Manifesto, 2007.

Futerra Sustainability Communications, The Greenwash Guide, 2008.

October 2008 report, Energy Efficiency, Innovation and Job Creation in California, by UC Berkeley economist David Roland-Holst

Co-Operative Investment WWF, Unconventional Oil: Scraping the Bottom of the Barrel, 2008.

Smith, K., of PLATFORM for BankTrack, Friends of the Earth – Scotland, People & Planet, Scottish Education and Action for Development, Stop Climate Chaos and PLATFORM, Cashing in on Coal. RBS, UK Banks and the Global Coal Industry, August 2008.

The Climate Group, The Business Guide to the Low Carbon Economy: California, 2008

The Climate Group, Briefing Note 2, 10 tips for developing your carbon footprint, March 2007.

The Climate Group, Briefing Note 3, 10 Things to consider … when disclosing carbon risks and benefits, April 2007.

Thaler, R. H., Sunstein, C. R., Penguin, Nudge: Improving Decisions about Health, Wealth and Happiness, 2008.

Smith, A., Baird, N., for Friends of the Earth, Harper Collins, Save cash & save the planet, 2005.

Forum for the Future, Check-out carbon – the role of carbon labelling in delivering a low-carbon shopping basket, July 2008.

Hopkins, R., Green Books, The Transition Handbook: from oil dependence to local resilience, 2008.

Boyde, T., New Economics Foundation, Cusgarne Organics Local Money Flows, Part of the Plugging the Leaks Programme, 2001.

Acronyms and abbreviations

CCS	Carbon Capture and Storage
CDM	Clean Development Mechanism
CER	Certified Emission Reduction
CO_2	Carbon dioxide
CO_2e	Carbon dioxide equivalent
COP	Conference of the Parties (to the UNFCCC)
DoE	Department of Energy (US)
EC	European Commission
EPA	Environmental Protection Agency (US)
EPRI	Earth Policy Institute
ERU	Emission Reduction Unit (traded under JI, worth one tCO_2)
ETS	Emission Trading Scheme
EU	European Union
EUA	EU Allowance (trading unit under the EU ETS, worth one tCO_2)
EU ETS	European Union Emission Trading Scheme
FAO	UN Food and Agricultural Organization
GDP	Gross Domestic Product
GFN	Global Footprint Network
GHG	Greenhouse Gas
GWh	Gigawatt hour
IEA	International Energy Administration
IFPRI	International Food Policy Research Institute
IPCC	Intergovernmental Panel on Climate Change
ITL	International Transaction Log (recording system for Kyoto transactions)
JI	Joint Implementation

LPI	Living Planet Index
LULUCF	Land-Use, Land-Use Change and Forestry
MGGA	Midwestern Greenhouse Gas Accord (US)
MW	Megawatt
MWh	Megawatt hour
NAP	National Allocation Plan (EU ETS)
OECD	Organization for Economic Co-Operation and Development
PV	Photovoltaic
REDD	Reduced Emissions from Deforestation and Degradation
RGGI	Regional Greenhouse Gas Initiative (US)
TWh	Terrawatt hour
UNCCD	United Nations Convention to Combat Desertification
UNEP	United Nations Environment Programme
UNFCCC	United Nations Framework Convention on Climate Change
USDA	United States Department of Agriculture
VCS	Voluntary Carbon Standard
VER	Verified Emission Reduction (traded on the voluntary market, worth one tCO_2)
WCI	Western Climate Initiative (US)
WTO	World Trade Organization
WWF	formerly known as the World Wildlife Fund, now WWF

Glossary

Adaptation – Actions undertaken to help communities and ecosystems to cope with the consequences of changing climate conditions, including raising coastal dikes, substitution of heat-resistant strains of plants for existing strains.

Additionality – An approval test for projects under the Clean Development Mechanism (CDM) – projects are considered additional if they would not have been developed in the absence of the CDM.

Allocation – The granting of emissions allowances to GHG emitters to establish an emission trading market.

Annex 1 – The 36 countries in the United Nations Framework Convention on Climate Change (UNFCCC) that have committed to a reduction in GHG emissions under the Kyoto Protocol, including all OECD countries and countries with economies in transition.

Annex II – The group of countries included in the UNFCCC, including all OECD countries in the year 1990, plus the European Union, and Central and Eastern European countries, except Albania and Yugoslavia. Under the Convention, these countries should provide financial resources to assist developing countries to comply with their obligations. Annex II countries are also expected to promote the transfer of environmentally sound technologies to developing countries.

Anthropogenic GHG emissions – The excess amounts of CO_2 released into the atmosphere by human activities, leading to rapid climate change.

Baseload Power – The minimum level of demand on an electrical supply system, also meaning the load that exists 24 hours a day. Baseload power plants provide energy at a constant rate, usually at a relatively low cost.

Biodiesel – A biofuel in which organically-derived oils (soybean or canola oils, animal fats, waste vegetable oils, or micro algae oils) are combined with alcohol and blended with conventional diesel fuel or used by itself.

Biodiversity – The variety of species that inhabit an area. Maintaining biodiversity helps to prevent species extinction and is important for the health of ecosystems.

Bioenergy – Energy derived from biomass.

Bioethanol – Liquid fuel generated from fermented sugar (alcohol). Bioethanol is made from starch plants (grain, mostly corn, and tubers like cassava); sugar beet or sugar cane.

Biofuel – Gas or liquid fuel made from organic material.

Biogas – Methane-rich gas produced through the fermentation of dung or crop residues in an airtight container.

Business As Usual Scenario (BAU) – A projection of future emission levels in the absence of changes in current policies, economics and technology.

Cap-and-Trade – An emissions trading system, where total emissions are limited or 'capped' and the credits traded between registered entities.

Carbon – Basic chemical element in all organic compounds. When combusted, it is transformed into CO_2 – one tonne of carbon generates about 2.5 tonnes of CO_2.

Carbon debt – The overuse of the CO_2 absorption capacity of parts of the environment. For example, when land is converted from peatland to biofuel the conversions can cause higher emissions than the annual savings from replacing fossil fuel with biofuel – this is the carbon debt that must be repaid before there will be any impact on cutting GHGs.

Carbon (dioxide) capture and storage (CCS) – A process consisting of the separation of CO_2 from industrial and energy-related sources, transport to a storage location, and long-term storage in oil wells or acquifers.

Carbon dioxide (CO_2) – A naturally occurring, colorless, odorless gas that is a normal part of the atmosphere, created through respiration, as well in the decay or combustion of animal and vegetable matter. Also a by–product of burning fossil-fuels from fossilized carbon deposits, such as oil, gas and coal, as well as some industrial processes. Because it traps heat radiated by the Earth into the atmosphere, it is called a GHG, and is a major factor in potential climate change.

Carbon dioxide equivalent (CO_2e) – A measure used to compare the emissions from various GHGs based upon their global warming potential (GWP). CO_2e are commonly expressed as million tonnes of carbon dioxide equivalents ($MTCO_2E$). One unit of CO_2e is equal to one metric tonne of carbon emissions.

Carbon emissions [also carbon dioxide emissions] – The release of CO_2 (or CO_2e gases) into the Earth's atmosphere.

Carbon footprint – The impact human activities have on the environment, in terms of the amount of GHG's produced, measured in units of CO_2.

Carbon neutral – An attempt to offset your carbon footprint. Once emissions have been reduced, the remainder is calculated and offsets are purchased to counteract the remainder.

Carbon sequestration – The process where CO_2 is absorbed in such a manner as to prevent its release into the atmosphere. It can be stored underground (see CCS) or stored in a carbon sink, such as a forest, the soil or in the ocean.

Carbon sink (also known as sinks) – Carbon reservoirs and conditions that absorb and store more carbon than they release. Forests and oceans are large natural carbon sinks.

Carbon tax – A tax by governments, usually on the use of carbon containing fuels, but can be a levy on all product associated carbon emissions.

Certified Emission Reduction (CER) – A credit or unit equal to one metric tonne of CO_2e, generated under the CDM.

Clean Development Mechanism (CDM) – This is the mechanism laid out in article 12 of the Kyoto Protocol, that allows Annex I parties to the Kyoto Protocol to make emissions reductions in Annex II countries.

Climate – Climate is not the same as weather. What this actually refers to is the statistical description in terms of the mean and variability of relevant quantities over a period of time, which can range from a period of months to millions of years. The classical period for averaging these variables is 30 years, as defined by the World Meteorological Organization (WMO). The relevant quantities are most often surface variables such as temperature, precipitation and wind.

Climate change – A term referring to a change in the state of the climate, used to imply a significant change from one climatic condition to another.

Combined Heat and Power (CHP) – Generates both electricity and heat (eg, hot water). The heat can then be used to heat homes or businesses, as long as there is a demand.

Community Independent Transaction Log (CITL) – Independent transaction log recording the issue, transfer and cancellation of carbon allowances within the European Union.

Compliance – Achievement by a Party of its quantified emission limitation and reduction commitments under the Kyoto Protocol.

Conference of the Parties (COP) – The supreme body of the UNFCCC.

Crediting Period – The duration when a project generates carbon credits, and cannot extend beyond the operational lifetime of the project

Deforestation – The removal or destruction of areas of forest, with no plans to restore it.

Desertification – Land degradation in arid, semi-arid and dry sub-humid areas resulting from various factors, including climatic variations and human activities.

Double Counting – When a transaction, or credit, is counted twice. This occurs when different stakeholders claim offsets through their involvement. At an international level, it would mean a project covered by the EU ETS also counting for the generation of Emission Reduction Units through Joint Implementation – which is not allowed.

Economies in Transition – Countries that are in the transition from a planned economy to a market-based economy, i.e. the Central and East European countries,' Russia, and the former republics of the Soviet Union.

Glossary

Ecosystem – A system of living organisms interacting with each other and their physical environment.

Energy – Often defined as the ability to do work. There are different forms of energy: heat, light, mechanical, electrical, chemical and nuclear. In the context of electricity rates and services, the word energy refers to electrical energy and is measured in watts per hour – power multiplied by time. In this sense, energy is a measure of the quantity of units of electricity used in a given time period, measured in kilowatt-hours (kWh).

Energy efficiency – The ratio of the useful output of services from an article of industrial equipment to the energy use by such an article; for example, the number of operational hours of a filling machine per kW hour used, or vehicle miles traveled per gallon of fuel (mpg).

Energy from Waste (EfW) – also known as Waste to Energy (WtE). Encompasses a range of technologies that directly generate energy from waste. It does not cover waste management techniques that save energy, such as recycling.

Environmental sustainability – Meeting the needs of the present without compromising the ability of future generations to meet their needs.

Emission Reduction – Monitored reductions in GHG emissions which would have occurred in the absence of a project as described on the project baseline.

Emission Reduction Unit (ERU) – A unit generated by a Joint Implementation project, equal to one metric tonne of CO_2e, issued under article 6 of the Kyoto Protocol.

Emission Reduction Purchase Agreement – Binding purchase agreement signed between buyer (of CERs or ERUs) and seller.

European Union Allowances EUA – A tradable unit under the EU ETS. Equals 1 tonne of CO_2.

European Union Emissions Trading Scheme (EU ETS) – part of the EU cap-and-trade scheme, used for the exchange of emissions permits.

External forcing – This refers to a forcing agent outside the climate system causing a change, for example, volcanic eruptions, solar variations and anthropogenic changes in the composition of the atmosphere and land-use change.

Food security – Secure physical and economic access to sufficient amounts of nutritious food required for normal dietary needs and growth.

Fossil-fuels – these contain carbon and were formed over millions of years through the decay, burial and compaction of rotting vegetation on land, and of marine organisms on the sea floor. The three main types of fossil-fuel are coal, oil and natural gas – the combustion of these fuels in generating power is the major way in which humans add to the GHGs in the atmosphere.

Fuel cell – An electrochemical engine (no moving parts) that converts the chemical energy of a fuel, such as hydrogen, and an oxidant, such as oxygen, directly to electricity.

Global carbon budget – The balance of exchanges (income and losses) of carbon between the carbon reservoirs or within one specific loop (e.g., atmosphere and biosphere) of the carbon cycle. An examination of the carbon budget of a pool or reservoir can provide information about whether it is functioning as a source or sink for CO_2.

Global warming potential (GWP) – The ratio of warming caused by a substance compared to warming caused by a similar mass of CO_2. The GWP of carbon dioxide is 1.0.

Globalization – The growing integration of countries worldwide through the increasing volume and variety of crossborder transactions in goods and services, free international capital flows, and the rapid and widespread diffusion of technology, information and culture.

Gold Standard (GS) – Accreditation awarded to CDM and voluntary projects that have higher sustainable development credentials than required by the CDM rules. The GS was set up by a group of environmental NGOs who wanted to encourage developers to run high quality projects. The GS board run a GS VER accreditation scheme for the voluntary market.

Greenhouse gases (GHG) – Gases which, when present in the atmosphere, slow the infrared radiation of heat from the Earth, upsetting the energy balance. They include water vapour, carbon dioxide (CO_2), methane (CH_4), nitrous oxide (N_2O), halogenated fluorocarbons (HCFCs), ozone (O_3), perfluorinated carbons (PFCs), and hydrofluorocarbons (HFCs).

Greenhouse gas emissions (GHG Emissions) – The GHGs we discharge into the air. The major emission adding to the greenhouse effect is CO_2, but other emissions, such as methane and nitrous oxide, absorb energy more efficiently than CO_2 and thus have a higher impact per amount emitted.

Greenhouse Gas Protocol (GHG Protocol) – The widely used international accounting tool for government and business leaders to understand, quantify, and manage GHG emissions.

Gross Domestic Product (GDP) – The monetary value of all goods and services produced within a nation.

Gross National Product (GNP) – The monetary value of all goods and services produced by a nation's economy, including income generated abroad by domestic residents, but without income generated by foreigners.

Group of Eight (G8) – An informal group of some of the world's largest economies, it was originally set up following the oil shocks and global economic recession of the 1970's. The Group consists of France, Germany, Italy, UK, Japan, Canada, the US and Russia. The leaders of each country meet every year to discuss how best to manage global economic challenges.

Group of Eight plus Five (G8+5) – This consists of the leaders of the G8 countries and the heads of five leading emerging economies: Brazil, China, India, Mexico and South Africa. Formed at the 2005 G8 summit in Gleneagles, Scotland to create a stronger global body, to accelerate movement on trade talks at Doha, as well as improve cooperation on climate change.

Group of 77 and China – China leads the developing country group in the UN climate negotiations. The group consists of more than 130 developing countries.

International Transaction Log (ITL) – A planned centralized database of all tradable credits under the Kyoto Protocol and the application that verifies all international transactions and their compliance with Kyoto rules and policies.

Intergovernmental Panel on Climate Change (IPCC) – established by WMO and the United Nations Environmental Programme (UNEP) in 1988 to review scientific, technical and socio-economic information relevant for the understanding of climate change, its potential impacts and options for adaptation and mitigation. It is open to all Members of the UN and of WMO and involves over 2,000 climate experts.

Joint Initiative – Mechanism of the Kyoto Protocol that allows Annex I parties to generate emissions reductions through the development of projects in other developed countries (such as Eastern Europe).

Joint Implementation (JI) – Joint Implementation is a mechanism for transfer of emissions permits from one Annex B country to another. JI generates ERUs on the basis of emission reduction projects leading to quantifiable emissions reductions.

Kilowatt – A unit of measure for electricity that equals one thousand watts.

Kilowatt-hour (kWh) – The amount of energy present in one kilowatt of electricity supplied for one hour of time. Electricity is sold as kilowatt-hours. People pay for electricity by the number of kilowatt-hours used.

Kyoto Protocol – Adopted in 1997 in Kyoto, Japan, at the Third Session of the COP to the UNFCCC. It contains legally binding commitments, in addition to those included in the UNFCCC. Countries included in Annex B of the Protocol (most Organization for Economic Cooperation and Development countries and countries with economies in transition) agreed to reduce their anthropogenic GHG emissions (carbon dioxide, methane, nitrous oxide, hydrofluorocarbons, perfluorocarbons, and sulphur hexafluoride) by at least 5% below 1990

Glossary

levels in the commitment period 2008 to 2012. The Kyoto Protocol entered into force on 16 February 2005.

Landfill – A large pit where solid waste is deposited, and buried under shallow layers of dirt.

Land-use and Land-use change – The term land-use is used in the sense of the social and economic purposes for which land is managed (e.g., grazing, timber removal, and conservation). Land-use change refers to a change in the use or management of land by humans.

Leakage – Decrease or increase of GHG-related benefits outside the boundaries set for defining a project's net GHG impacts that result from project activities.

Major Economies Meeting on Energy and Climate Change (MEM) – Launched by US President Bush, in May 2007, to develop and contribute to a post-Kyoto framework on energy security and climate change by the end of 2008.

Marginal Abatement Cost (MAC) – In the context of the carbon market, this is the cost of reducing emissions by one tonne of CO_2. Aggregated marginal costs over a number of projects or activities defines the marginal abatement cost curve.

Member state – A state which is a member of an international organization or group.

Meeting of Parties (MOP) – Equivalent to the COP, but the terminology differs between agreements. There is a tendency to use the term COP with reference to the conventions, and Meeting of the Parties for protocols. The first MOP to the Kyoto Protocol was held in Montreal in December 2005 during the 11th COP.

Mitigation – Human intervention to reduce the sources of or increase the sinks of GHGs.

Monitoring – The collection and archiving of all relevant data necessary for measuring anthropogenic emissions by sources of GHGs within the project boundary of a project activity and leakage, as applicable.

National Allocation Plan (NAP) – Allocation of emission allowances at the national level to individual sites under EU ETS.

Non-governmental organization (NGO) – A non-profit group organized outside of political structures to realize particular social and/or environmental objectives.

Non-Annex I countries – Developing countries, with no binding emission reduction targets, under the Kyoto Protocol.

Non-renewable energy – Energy from sources that, when consumed, cannot be replaced or renewed. Sometimes used as another term for fossil-fuel.

Offset – The purchase of a carbon credit to offset emissions being made, either by a government, corporation or individual.

Oil sands and oil shale – Sands, sandstone rock and shales containing bituminous material that can be mined and converted to a liquid fuel. This transformation is highly energy and water intensive, and environmentally destructive.

Passive energy saving – Term referring to the technologies which minimize the use of energy, in industry or consumer electronics.

Peatlands – Wetlands where the soil is highly organic because is it formed mostly from incompletely decomposed plants. This soil is called peat and its presence is what defines peatlands.

Photovoltaic cells – Cells, usually made of specially-treated silicon, that transform solar energy from the sun to electrical energy.

Polluter-pays principle – The idea that the person that causes pollution should pay to put right the damage that it causes.

Programmatic CDM Projects – Activities to reduce emissions such as implementation of a government measures or private sector initiatives.

Radiative Forcing – A change in the balance between incoming solar radiation and outgoing infrared radiation. Without any radiative forcing, solar radiation coming to the Earth would continue to be approximately equal to the infrared radiation emitted from the Earth. The addition of GHGs traps an increased fraction of the infrared radiation, radiating it back toward the surface and creating a warming influence.

Rapid climate change – Sometimes called

abrupt events or even surprises, such as a dramatic reorganization of the thermohaline circulation, rapid deglaciation, or massive melting of permafrost leading to fast changes in the carbon cycle.

Reforestation – Planting of forests on lands that have previously contained forests but that have been converted to some other use.

Regional Greenhouse Gas Initiative (RGGI) – A cooperative effort by 10 Northeastern and Mid-Atlantic states of the US to reduce CO_2 emissions by establishing a regional cap-and-trade programme. The aim is to cap and then reduce emissions by 10% by 2018.

Renewable or Biogenic CO_2 – CO_2 which is produced by burning carbon sourced from natural renewable materials, such as food waste or paper. Biogenic CO_2 is usually treated as having no impact on climate change, as it is part of the natural carbon cycle.

Renewable Energy – Energy obtained from sources that are inexhaustible, including wood, geothermal, wind, and solar thermal energy.

Reservoir – A component of the climate system, which has the capacity to store, accumulate or release a substance of concern, e.g. carbon, a GHG or a precursor. Oceans, soils, and forests are examples of reservoirs of carbon.

Resources – Raw materials, supplies, factories, offices, labour, management, and entrepreneurial skills that are used in producing goods and services. Or the substances that support life, including air, land, water, minerals, fossil fuels, forests and sunlight.

Sequestration – Carbon storage in terrestrial or marine reservoirs. Biological sequestration includes direct removal of CO_2 from the atmosphere through land-use change, afforestation, reforestation, carbon storage in landfills and practices that enhance soil carbon in agriculture.

Sustainable – Actions and/or products that meet our current needs without sacrificing the ability of future generations to meet their needs.

Sustainable development – Economic development that takes full account of the environmental consequences and is based on the use of resources that can be replaced.

Tipping point – the point at which slow, reversible change becomes irreversible. In climate terms it also refers to sudden or rapid change occurs.

Uncertainty – An expression of the degree to which a value (e.g. the future state of the climate system) is unknown.

United Nations Framework Convention on Climate Change (UNFCCC) – Adopted on 9 May 1992 in New York and signed at the 1992 Earth Summit in Rio de Janeiro by more than 150 countries and the European Community, for the 'stabilization of greenhouse gas concentrations in the atmosphere at a level that would prevent dangerous anthropogenic interference with the climate system'.

Urban heat island – The phenomenon of urban centres being warmer than the surrounding area. This effect is a product of the heat that is generated by all of the buildings present in urban areas and of their ability to absorb solar energy and release it as detectable heat.

Urbanization – The conversion of land from its natural state to cities.

Validation – The process of independent evaluation of a CDM project by an accredited Independent Entity according to requirements to CDM projects.

Verification – In order for CDM projects to have a formalized validation of an emission reduction stream, a recognized independent third party must confirm that claimed emissions reduction activity has occurred.

Verified Emission Reductions (VERs) – Generated by small scale projects, which are assessed and verified by third party organizations rather than through the UNFCCC.

Voluntary Market – Buyers and sellers of VERs that seek to manage their emission exposure for non-regulatory purposes.

Water security – Reliable availability of water in sufficient quantity to sustain human health, livelihoods, production and the environment.

Index

A

additionality 156–7
ARRO hierarchy 63, 64
auctions, carbon allowance
    159, 160, 179, 184, 185, 235,
    237
aviation 140–3, 168, 178, 196,
    197, 236

B

banking industry 213–15
Barclays 214–15
behavioural change 20–1,
    63–6, 128, 171, 196–7, 220–3
biocapacity 34–5
biodiversity losses 19, 35–6,
    145–6
biofuels 45, 46, 57, 132–7
biogas 102
biomass 101–4, 110, 111
blackouts 107, 108,, 112, 188
Brazil 35, 197–8
buildings energy
    management 124–9, 194,
    198–9

C

cap-and-trade 65–6, 69–71, 78,
    157–61, 164, 182–6
carbon capture and storage
    (CCS) 87–91, 214–5
carbon dioxide equivalent
    (CO_2e) 16, 18
carbon footprint 74–6, 183,
    195–6, 217–8, 223
carbon leakage 67, 71, 161, 169
carbon markets 30, 161–3,
    234–9
    see also carbon trading
carbon neutrality 207, 210
carbon price 86
    reasons for 25–7

setting 66–73
Carbon Reduction
    Commitment (CRC) 178–9
carbon sinks 19–20, 144–5
carbon taxation 66–72
carbon trading 13–14, 59, 65–6,
    69–73, 75, 141–3, 157–61,
    178–9, 182–6, 214
    see also carbon markets
Certified Emission
    Reductions (CERs) 154, 155,
    157, 172, 214
China 186–91, 201–2
cities
    climate change initiatives
        30, 191–201
    coastal cities 39, 191
    low-carbon design 192–4
    temperatures 127
Clean Development
    Mechanism (CDM) 152–7,
    162–3, 166–7
Clean Energy 2030 83–4
climate change
    causes of 15–18
    effects of 16–20
    speed of 16–17
climate change initiatives 80–3
    economic cost 22–5, 56–61
    urgency of 10, 21, 234
    see also cities; corporations;
    countries and individuals
clothing industry 51, 223
coal industry 87–91, 189,
    213–15,
command and control
    policies 64–5
compliance periods 69
conflict over resources 33–4,
    38–40, 45, 49
Congo Basin Forest Fund
    (CBFF) 150

consumer choices 220–9
contraction and convergence
    169–71
corporate social
    responsibility (CSR) 202
corporations
    climate change initiatives
        30–1, 56–60, 201–19
    shareholder pressure
        59–60, 142, 202
cost of tackling climate
    change 22–5
countries
    climate change initiatives
        29–30, 167, 175–91
crops
    failure see food scarcity/
        prices
    increased yields 47–8
Cuba 230
cultural change 12, 228–9

D

deforestation 16, 24, 35, 57,
    147–50
    and biofuel production 134
desertification 33, 40, 44
droughts 24, 28, 33, 38, 39, 46,
    98
Dubai 192–3

E

ecological footprints 34–5
    see also carbon footprint
    and water footprint
electric vehicles 80, 131,
    137–40, 207–8
electricity industry 80
electricity prices 68
embedded carbon 74–5, 77,
    168–9, 171–2, 223
energy consumption 21, 81

energy efficiency 118–29,
        188–9, 197, 199–200
    appliances 203
    building retrofits 194–5
    corporations 204–9
    hotels 208–9
    new builds 198–9
energy industry
    generation 115–17
    transmission and
        distribution 107–17
    see also electricity industry
        and renewable energy
Energy [R]evolution 83–4
energy security 40, 42, 187
environmental valuation 52–4,
        149–50
equitable change to low-
        carbon 12, 27–31, 169, 239
    and international
        agreements 165–7
European Union
    Emissions Trading Scheme
        (ETS) 59, 65, 70, 75,
        141–3, 157–61
export-driven emissions 74,
        169, 190

F
finance, local 231
financial crisis, global 24, 26,
        61–2
flooding 19, 24, 28, 37–9
food scarcity/prices 18, 34, 44–
        8, 133–4, 230–1
forests 154
    and ecosystem regulation
        144–7
    valuation of 53–4, 149–50,
        236
fossil fuels 40
    scarcity/prices 42–4
    see also coal industry

and oil
fuel efficiency
    road vehicles 131–2
fuel sources, low-carbon
        84–107

G
geothermal power 83, 92,
        99–101
Google 83, 100, 207–8
Green New Deal 61–4
greenhouse effect 16
greenhouse gas emissions
    increase 19
    measurement 73–8
    reduction/strategies 20–2,
        56–61, 65–73, 79–150,
        151–7, 157–63, 167–72
    responsibility for 29–31,
        71–3, 152, 228–9
    stabilization 80–1
greenhouse gases (GHG) 15,
        16
greenwash 209–19, 227–9

H
heatwaves 24, 37
high-voltage direct current
        (HVDC) transmission 109,
        111–12
hurricanes 37–8
hydropower 84, 96, 98–9, 110

I
ice melt 17, 18, 44, 47, 49
individuals
    climate change initiatives
        30, 220–32
    pressure on corporations
        59–60, 142, 202
    see also behavioural
        change and consumer
        choices

industrial energy efficiency
        120–1, 168–9
Industrial Revolution 16, 145
Innocent 217–18
InterfaceFlor 215–17
Intergovernmental Panel on
        Climate Change (IPCC) 15,
        17, 18, 19, 37, 39, 140, 145,
        237
international agreements 27,
        165–7
    see also Kyoto Protocol
IT industry 113–15, 121–4

K
Kyoto Protocol 22, 29, 30, 69,
        73, 151–7, 164–5

L
labelling 77, 203, 225–6
Living Planet Index (LPI) 35
London, UK 64, 195–7, 208–9,
        221–2

M
Major Economies Meeting
        (MEM) 167
Masdar City, Dubai 193
measurement of carbon 73–8

N
negawatts 117, 121
nuclear power 85–7, 189

O
ocean acidity 17, 18, 144
offsetting 13, 63, 152, 156, 157,
        161–3,
oil
    supply peaks 42–4
    uses of 40–1
One Planet Living principles
        193

Index

P

peak oil 42–4
population, human 18, 29
    displacement/migration 33, 34, 38, 39
    urban concentrations 191
poverty 28, 29, 39, 41, 147, 166, 187

R

recycling 193, 206, 216, 218, 219
Reduced Emissions for Degradation and Deforestation (REDD) 147–9
renewable energy 92, 155, 159, 160, 188, 191, 197
    incentives 63, 177, 178
    intermittency 104–7, 110
    storage 104–7
    see also biogas; biomass; geothermal power; hydropower; solar power; tidal power; waste energy generation; wave power and wind power
resources
    conflict over 33, 38–40, 49
    ownership 28
risk management 55–6
Rizhao, China
    solar energy projects 200–1
road vehicles 130
    biofuels 134–6
    fuel efficiency 131–2

S

San Francisco, US 139
    energy-efficient new builds 198–9
São Paulo, Brazil
    waste energy

generation 197–8
Savoy Hotel 208–9
sea levels 17, 19, 24, 33, 39, 191
Seattle, US
    energy conservation 199–200
shipping 129, 130, 143–4, 168, 178, 236
Smart Grid 112–17
solar power 80, 83, 92, 93, 95–7, 108, 110, 113, 189, 200–1, 207, 212
species decline 18, 19, 35–6, 146
stabilization wedges 80
standards
    carbon measurement 76–8
    voluntary carbon offset credits 161–2
    see also labelling
Stern Review on the Economics of Climate Change 19, 22
Super Grid 108–11

T

taxation *see* carbon taxation
temperatures 15, 17, 18–19, 22, 47, 127
Tetra Pak 206
tidal power 98–9
transition to low-carbon 31–2
Transition Town movement 229–32
transport 79, 129–44
    freight 45
    public 84, 197
    sustainable 193

U

United Kingdom (UK)
    climate change initiatives/policies 176–80, 195–7
United Nations (UN) 38–9
    conferences (Bali; 2007) 54,

174; (Copenhagan; 2009) 174, 235, 236–9
United States (US) 198–200
    climate change initiatives/policies 36, 41–2, 62, 174, 180–6, 238

V

valuation, environmental
    forests 53–4, 149–50, 236
Vattenfall Abatement Curve 56–7, 91
virtual servers 123

W

Wal-Mart 204–5
waste diversion 193
waste energy generation 101–4, 197–8
water
    and manufacturing 52
    recycling 126
    and renewable energy 98–9, 99–100
    salination 39, 40
    scarcity 18, 19, 33, 38, 40, 44, 48–52
water footprint 51
wave power 98–9
weather, extreme 18, 19
    see also droughts; flooding, heatwaves and hurricanes
wind power 80, 83, 84, 85, 92–5, 105, 108, 110, 188, 189, 191